HUMANS, ANIMALS, AND U.S. SOCIETY IN THE LONG NINETEENTH CENTURY

HUMANS, ANIMALS, AND U.S. SOCIETY IN THE LONG NINETEENTH CENTURY

A Documentary History

Edited by
Dominik Ohrem

Volume I
Animal and Human in American Thought (Part 1)

Routledge
Taylor & Francis Group
LONDON AND NEW YORK

First published 2026
by Routledge
4 Park Square, Milton Park, Abingdon, Oxon OX14 4RN

and by Routledge
605 Third Avenue, New York, NY 10158

Routledge is an imprint of the Taylor & Francis Group, an informa business

British Library Cataloguing-in-Publication Data
A catalogue record for this book is available from the British Library

Library of Congress Cataloging-in-Publication Data
Names: Ohrem, Dominik editor
Title: Humans, animals, and U.S. society in the long nineteenth century : a documentary history / edited by Dominik Ohrem.
Description: Abingdon, Oxon ; New York : Routledge, 2025. | Includes bibliographical references and index. | Contents: v. 1. Animal and human in American thought (part 1) — v. 2. Animal and human in American thought (part 2) — v. 3. Domesticated and companion animals (part 1) — v. 4. Domesticated and companion animals (part 2) — v. 5. Wild animals.
Identifiers: LCCN 2025003165 (print) | LCCN 2025003166 (ebook) | ISBN 9780367470104 set | ISBN 9780367470005 v. 1 hardback | ISBN 9780367470029 v. 2 hardback | ISBN 9780367470036 v. 3 hardback | ISBN 9780367470098 v. 4 hardback | ISBN 9780367470104 v. 5 hardback | ISBN 9781003032717 v. 1 ebook | ISBN 9781003032748 v. 2 ebook | ISBN 9781003032779 v. 3 ebook | ISBN 9781003032809 v. 4 ebook | ISBN 9781003032816 v. 5 ebook
Subjects: LCSH: Human-animal relationships—United States—History—19th century—Sources
Classification: LCC QL85 .H8715 2025 (print) | LCC QL85 (ebook)
LC record available at https://lccn.loc.gov/2025003165
LC ebook record available at https://lccn.loc.gov/2025003166

ISBN: 978-0-367-46996-2 (set)
ISBN: 978-0-367-47000-5 hbk (vol I)
ISBN: 978-1-003-03271-7 ebk (vol I)

DOI: 10.4324/9781003032717

Typeset in Times New Roman
by Apex CoVantage, LLC

CONTENTS

GENERAL INTRODUCTION

According to historian Jürgen Osterhammel, the nineteenth century brought about no less than "the transformation of the world."[1] In discussing the enormous changes wrought during the period in an eponymous work of global history first published in German in 2009, Osterhammel stretches the nineteenth century backward and forward in time, eschewing "a sealed-off, self-sufficient history *of* the nineteenth century" in favour of "the insertion of an age within longer timelines: the nineteenth century *in* history."[2] Like British historian C. A. Bayly before him in *The Birth of the Modern World* (2004), Osterhammel thus adopts a "long" perspective on the nineteenth century to highlight connections and continuities that might otherwise be obscured by the traditional compartmentalisation of historical time into sharply edged uniform chunks.[3] Yet even in its regular shape, the nineteenth century was not only an exceptionally busy century characterised by several multi-level transformations that occurred as national variations of wider transatlantic or even global developments, it also highlighted more forcefully than others the immense consequences of (what has often been overgeneralised as) "human action" in a period now frequently invoked in debates about the epoch – or event – of the Anthropocene.[4]

The history of the United States offers an instructive window into the broader socioeconomic, cultural, and ecological developments of the long nineteenth century, such as the twin phenomena of industrialisation and urbanisation, while following a distinct path in other respects. What historians have called the "market revolution" heralded the transition from a largely community-centred and subsistence-oriented to a modern capitalist economy with a national market that itself became integrated into an expanding system of international commerce.[5] In the antebellum period, a significant upsurge in the extraction of bituminous and anthracite coal marked the rise of "fossil capital" and the first transition to high-energy-density fossil fuels, then primarily used in iron production and as fuel for the new steam engines, which in turn moved steamboats, locomotives, printing presses, powered looms, and other machinery. A more drastic rise in coal use occurred from the post-Civil War decades to the 1910s, which saw the beginning of a second energy transition from coal to oil. In the second half of the century, American manufacturing increasingly shifted to large-scale mechanised production in a growing factory system seeking to profit from the market and transportation advantages and concentrated labour force found in large cities, whose rapid growth was boosted by the predominantly urban-directed mass immigration taking place between 1880 and 1920. A rural society with an urbanisation rate of a mere five percent in 1790, by 1920 the U.S. had reached urban-majority status.[6]

Technological innovations in an industrialising society were accompanied by similarly momentous developments in the modern sciences, which gradually emerged as specialised and differentiated domains of knowledge production. For much of the nineteenth century, scientific and technological knowledge are best understood as relatively distinct epistemic domains that were not defined by a consistent hierarchical or causal relationship (or the kind of quasi-symbiosis evoked by the term "technoscience"), as innovations like steam technology and textile machinery were driven by practical experimentation rather than relying on application of formal scientific theories.[7] The developing electrical industry, on the other hand, was heavily dependent on research in electromagnetism, and the awareness of the importance of science led to the creation of dedicated industrial research laboratories like Thomas A. Edison's Menlo Park laboratory in New Jersey (established in 1876), which would play an important research and development role in an increasingly competitive corporate landscape.[8] In the field of medicine, the creation of vaccines as a biotechnology was facilitated by knowledge about microbial disease origins and transmission established by the germ theory of disease, leading to the rapid development of rabies, cholera, typhoid, and plague vaccines nearly a century after Edward Jenner's pioneering smallpox vaccine. As is underlined by the long-nineteenth-century arc of evolutionary theory from Jean-Baptiste Lamarck to the decades-long debates around Charles Darwin's theory (especially its key concept of natural selection) to the latter's synthesis with Gregor Mendel's rediscovered work on inheritance, scientific developments could of course also have enormous social impact even if they did not lend themselves to any purposeful practical application.[9]

The outsized importance of territorial expansion and slavery, phenomena that became explosively intertwined in the leadup to the Civil War, distinguished the U.S. from European countries during the period. Between its founding and the beginning of World War I, the U.S. developed from a fledgling settler nation into a global power that ruthlessly realised its territorial ambitions in North America at the expense of European imperial powers (a process largely completed by 1850) and Native peoples (a process largely completed by 1890), established its claim to hegemony over affairs in the western hemisphere, and transitioned from a continental to an "offshore" phase of empire-building with the annexation of Hawaii and the acquisition of Puerto Rico, Guam, and the Philippines through Spanish cession at the conclusion of the Spanish-American War (1898). The global cotton boom that began around 1800 fuelled not only the southern plantation economy but also a growing northern textile industry with its ceaseless demand for raw cotton, integrating both sections into a complex world economy in which the profits of slavery-driven agriculture fed the engines of industrial capitalism. Boosting the domestic trade in enslaved people, the rise of the "cotton kingdom" and its subsequent expansion to and beyond the Mississippi were unmistakable indicators that slavery would not simply disappear on its own, as earlier "gradualists" like George Washington and Thomas Jefferson had suggested.[10]

Although varying in their impact and ability to engender long-term structural change, the emergence of social movements – distinct from previous forms of collective (protest) activities in their level of organisation and structure, ideological grounding and commitment, and impetus for systemic change – was another important development among the profound social transformations of the period. And here, too, formal and informal transatlantic relationships consisting in the transfer of ideas, strategic collaborations, and/or international conferences and movement organisations played a significant role. Both in its origins in Quaker networks in the second half of the eighteenth century and in its later organised form in the antebellum period, antislavery activism provides an early prominent example for the transnational cooperation that would characterise many subsequent social movements. And beyond its transatlantic dimension, antislavery activism was also defined by a degree of intermovement collaboration with other contemporary forms of social activism, including the women's rights movement.[11]

As even this very cursory overview suggests, the history of the U.S. in the long nineteenth century makes for some compelling historical drama. It is the story of a group of colonies that shook off an imperial overlord only to become an aggressive imperial power in its own right; the story of a society in the throes of a series of far-reaching transformations and a sweltering internal conflict that threatened to tear it apart; the story of a society inexorably hurtling towards modernity. It is also a story that, like most others of its kind, has traditionally been told as an exclusively *human* story for reasons that have much to do with the sturdy conceptual bond between humanity and historicity coming out of the European tradition. The doctrine that "[a]ll History must necessarily have reference to the existence and condition of *Man*" articulated by former president John Quincy Adams in an 1845 article was indeed a truth so self-evident that it hardly required explication.[12] Nor was the figure of "Man" – understood to be the only truly historical being and agent of history – defined solely in contradistinction to nonhuman animals and nature; it also excluded those supposedly less-than- or infrahuman peoples whom settler discourse construed as caught in a state of quasi-ahistorical stasis due to their supposed lack of civilisational "progress."

In focusing on the role of human-animal relations – broadly understood here as involving both the plethora of embodied social relations between human and nonhuman animals and, relatedly, ideas about "the human" and "the animal" as (distinct, overlapping, hierarchical, symbiotic, etc.) ontological domains – the present collection follows a now already substantial line of animal historical scholarship that has put to scrutiny foundational principles of the discipline by challenging, and pushing beyond, its anthropocentric delimitations. Not unlike, for example, postcolonial or feminist critiques of traditional historical practice, the perspectives of historical animal studies have posed fundamental questions about the underlying premises and assumptions this practice has been based on at least in its European(-derived) manifestation. This includes asking who, under what conditions, can become a subject of history, is endowed with a historicity that also underwrites their subjectivity and perhaps even their ethical considerability? Rather than exhausting itself in a long list of new or unduly neglected topics, the field of historical animal studies has established a new optics with which "to reread the historical constitution of animal-human societies," a lens that also allows insights into the ways in which ideas about animality and particular types of animals have informed human intergroup relations (and vice versa) through a constitutive interplay with social vectors like race and gender.[13]

The five volumes of *Humans, Animals, and U.S. Society in the Long-Nineteenth Century* cover the roughly 140 years from the early national period following the American Revolutionary War (1775–1783) to the end of the Progressive Era around 1920, documenting the evolving landscape of human-animal relations in U.S. society – and, to a lesser extent, on the North American continent more broadly – during a period of profound socioeconomic, technological, and ecological change. Taken together, the documents collected here testify to the ubiquitous presence of animals across the entire fabric of American life in multiple and sometimes overlapping roles. Animals were exploited as workers in various capacities: as providers of essential agricultural services and products; as crucial means of transportation in rural and urban, commercial, industrial, and military settings; as exhibits or performers for human audiences in increasingly popular forms of public display or spectacle. They were cherished as companions inside human homes (and sometimes in less ordinary places). Some were celebrated as heroes for saving human lives or for accomplishing extraordinary athletic feats, while others were hunted for food, sport, or profit, vilified as vicious killers, "criminals," or enemies of "civilisation," or suffered as experimental subjects in physiological laboratories and university lecture halls. Some assumed vital complementary or even enabling roles in relation to the sweeping transformations of the long nineteenth century, while others resisted these changes or persisted in spite of them. Many, including the numerous species and populations of wild animals who found themselves confined to ever-shrinking habitats in the face of American territorial and economic expansion, were quite unambiguously on the receiving end of these transformations, while others were able to adapt to them or even flourished in the synanthropic (including synurbic) niches created in increasingly human-altered and human-dominated ecosystems or "anthromes."[14]

ANIMALS, SOCIETY, AND HISTORICAL CHANGE

As is the case with other times and other places, animals' lives and experiences in the long nineteenth century varied greatly depending on their respective positions in, or in relation to, human society. These positions are best understood as transindividual or *structural* in nature, tied to the enduring patterns and logics of social life commonly discussed under the rubrum of "social structure."[15] Reflecting inherited anthropocentric ideas about society, this concept has traditionally been limited in its adoption by historians to the configurations of enabling and constraining conditions that have shaped *human* life and agency in specific times and places. But if we follow the aforementioned tenet that human social worlds have always constitutively involved relations with other species, and should thus be understood as fundamentally more-than-human, this raises the question of how animals, too, have been impacted by dominant regularities of social life that are reflected in everyday modalities of human-animal interaction and have crystallised into various animal-centred institutions.

According to historian William H. Sewell, social structures can be understood as composites of "cultural schemas" (comprising "meanings, motivations, and recipes for social action"), "distributions of resources," and "modes of power" (determining the relative legitimacy of cultural schemas and regulating access to resources).[16] A key feature of social structures, their semi-autonomy, manifests in their relationship of asymmetric reciprocity with individuals and collectives – their power to shape the latter in significant ways while themselves being fairly resistant, though not irresponsive, to transformative individual or collective agency, allowing for "patterns of relations to be reproduced, even when actors engaging in the relations are unaware of the patterns or do not desire their reproduction."[17] An important distinguishing criterion for the analysis of social structures suggested by Sewell is their "depth." Deep structures are generative of surface structures and exhibit a much higher degree of resistance to structural change than the latter, are characterised by their pervasive influence on a "wide range of institutional spheres, practices, and discourses," and, as largely "taken-for-granted mental assumptions or modes of procedure," tend to elude conscious access.[18] Deep structures may thus be understood as "structures of structures," creating vertical dependencies with more mutable surface structures and commonly following distinctly longue-durée temporalities of historical change. Deep structural persistence might also be facilitated by horizontal integration, that is, by relationships between different deep structures characterised by mutual reinforcement and by a co-shaping of surface structures. Patriarchy, for example, would certainly fit the description of such a diachronically persistent and synchronically pervasive deep structure – as would what sociologist Erika Cudworth has called anthroparchy, denoting the "complex and relatively stable set of hierarchical relationships in which 'nature' is dominated through formations of social organization which privilege the human."[19] Structural perspectives are thus useful for analysing the historical (trans)formation of social relations of power like those between humans and animals and their intersection with other entrenched structural power dynamics ("anthropatriarchy" might serve as a heuristic descriptor for the structural connection between human and male supremacy that has shaped social relations in European or European-derived settler societies like the United States).

Underscoring the (staying) power of social structures does not mean ascribing to them quasi-dictatorial authority and/or a near-imperviousness to change. Nor is agency – a term that has been more intensely debated in its relation and applicability to animals than that of structure[20] – to be understood as some external force whose ability to produce or promote social change is premised on "freedom" from structural impositions. The durability of social structures hinges on their ability to *harness* agency by channelling it towards the structurally reproductive rather than the structurally subversive. The relationship between structure and agency is thus one of co-implication, and the possibility for change inherent to the very functioning of structures, not least because their multiplicity and intersectionality and the transposability of schemas to a variety of contexts mean that social practice can produce unpredictable results with transformative impulses. For Sewell, historical change is particularly likely when events produce structural uncertainties by causing schemas, resources, and modes of power to no longer align and coalesce "into self-reproducing streams of structured social action," allowing for "dislocations and transformative rearticulations of structure."[21]

Emphasising the importance of social structures in the analysis of human-animal relations does not mean that animal lives are pre-determined in a way that would essentially restrict any form of human-animal relationship to a fixed set of possible (inter)actions. What it means, however, is that especially (but not only) domesticated animals are, as Bob Carter and Nickie Charles put it, "agents in relation to human dominated structures."[22] Even as animal-centred institutions in Europe and its former settler colonies (and elsewhere) may involve fairly disparate human-animal practices and reflect different cultural, economic, and ethical perspectives on human-animal relations, the majority of them are characterised by a shared consensus about the legitimacy of human sovereignty, a fundamental structuring principle that has served to normalise the near-absolute control over animals' lives, including their movements and uses, their exposure to or protection from violence, and even their morphological, physiological, and behavioural characteristics.

When thinking about potentials for change, however, it is crucial to consider that the structural circumscription of animal life occurs not only through direct mechanisms of control and coercion at work in different institutional contexts but also, in a mediated yet powerful sense, through the humans interacting with them: these, too, are "agents in relation to human dominated structures," guided in their behaviour as individuals by the self-reproducing character of institutionalised social and economic practices, which creates difficulties for thinking and acting otherwise, against or around dominant epistemic and normative frameworks of human-animal relations. The long-nineteenth-century trajectory of animal advocacy from early relatively isolated individual voices lacking broad societal support to the institutionalisation of animal advocacy through an organised animal welfare movement wielding a modicum of legislative and policy influence, provides an example of structural change fostered by (in Sewell's terms) the transposition of cultural schemas about, among other things, the socially deleterious effects of cruelty and the alleviation of suffering as a moral imperative for/in "civilised" society to the context of human-animal relations. At the same time, even the later more institutionalised forms of nineteenth-century animal advocacy operated within the epistemic and material confines of an anthroparchal deep structure that enabled it to challenge the legitimacy of certain manifestations of human sovereignty without challenging the principle itself.

As John Berger has argued in a well-known 1980 essay, the enormous structural transformations of the long nineteenth century, most of which were directly or indirectly tied to the development of industrial capitalism, heralded a modern separation between humans and animals through an ever more absolute marginalisation of the latter under a rapidly growing regime of capitalist economic and social relations, prefiguring the "new solitude" experienced by the human denizens of post-industrial societies deprived of any meaningful connection to their unspeaking companions.[23] For Berger, this connection, which had predated the transformations of "the last two centuries" and been exemplified by the rhythms of traditional agrarian life and labour, was lost, to be replaced by the (non-)relationship between the alienated "urban stranger" and the ensemble of marginalised and thoroughly "deanimalised" creatures now increasingly populating family homes or zoos – creatures whose diminished existence mirrored humans' own reduction "to isolated productive and consuming units" under industrial and corporate capitalism.[24]

Berger's essay has been criticised for its lack of historical precision and overly linear account of historical transformation, which tends to flatten a more complex reality in which human-animal relations could be subject to quite different temporalities of change, and for its anthropocentric framing of these relations, which for Berger matter mostly in what they mean for the human side of the equation.[25] Urbanisation and industrialisation are clearly the main culprits in Berger's diagnosis of the irreversible "disappearance" of animals, echoing a broader perspective according to which the explosive growth of cities (regarded as quintessentially "human" spaces) and the rise of mechanised labour (seen as replacing "pre-industrial" animal power) encapsulated the dual process of alienation from and mastery over nonhuman life characteristic of the period.

As one might expect, historical reality was messier and less linear. Even though nineteenth-century cities were unmistakably anthropogenic spaces, this did not mean that they were also anthropoexclusive. Instead, these burgeoning urban environments were shaped by and home to a wide range of nonhuman organisms, all of which played a role in the ecological dynamics and social experience of city life. In fact, it was precisely in reaction to the nature of cities as often unruly multispecies landscapes that various attempts at removing or regulating urban animal presences were put in place. And while the long nineteenth century was shaped by a gradual transition from an organic to a mineral energy regime, industrial and animal power existed in a crucial co-enabling relationship, as new technologies relied upon and sometimes actively expanded their older but rarely anachronistic animal-powered counterparts. Often associated with rapid or even revolutionary rather than incrementalist transformations, industrialisation was not simply a story of animal disappearance but, as historian Edmund Russell explains, "a biological as well as a mechanical process" that "never came close to replacing living organisms completely."[26] Traditional animal traction remained indispensable, as industrial production depended on a still largely animal-powered transportation system in which humans and animals continued to interact daily. Rather than indicating the decline of animal transportation, the expansion of the railroad network, for example, caused the American equine population to grow, as animal power was needed almost everywhere to haul passengers and cargo from and to the railroads or bridge various unconnected rail lines. Similarly, the Civil War, widely understood as the first "industrial war," with its factory-produced and railway-distributed weaponry and ammunition, relied on an enormous number of horses and mules to put the new martial technologies to use on the battlefield. Until well into the twentieth century, coal mining, a main driver of the industrial revolution, relied on the labour of mules in mine haulage systems (although ponies and even dogs were also used) in mining states like Pennsylvania, West Virginia, or Wyoming, where the animals moved heavy coal carts or delivered supplies to their human co-workers underground. And although new industrial technology like the mechanical reaper significantly increased agricultural efficiency, steam power – costly and often impracticable – never managed to replace animal traction on a large scale. In short, then, equating urbanisation and industrialisation with the "disappearance" of animals distracts from the essential complementary roles they fulfilled – including as distinctly urban and distinctly industrial workers – throughout the entirety of the long nineteenth century.[27]

Imprecise as it may be in terms of historical analysis, however, Berger's essay still manages to capture a range of enormously consequential structural transformations in human-animal relations that collectively amounted to a new regime of human sovereignty, one that was (and still is today) characterised by an unprecedented "commodi-faunation" – or commodification of animal life – in and across various socioeconomic and cultural contexts.[28] Selective breeding, industrialised slaughter, the international animal trade, and the developing pet industry exemplify the reach of this new regime and its multifarious institutional forms. The subjugation of animals by this regime is perhaps most readily associated with the millions of domesticated animals living and dying within elaborate systems of exploitation whose more blatantly violent aspects became largely removed from the experiential horizons of everyone not directly involved in their operation, allowing them to function "almost imperceptibly within everyday peaceable human relations."[29]

Most evident in the case of the ruthless machinery of a growing meatpacking industry, this manufactured imperceptibility was itself a product of the convergence of a number of nineteenth-century developments, including the identification of the waste products of urban slaughterhouses and rendering plants as a severe public health issue, anti-cruelty sensibilities that led to and were further amplified by the first organised animal advocacy movement, and the sheer space and infrastructure requirements of an industry that "processed" untold numbers of animals per day to satisfy the market demands produced by growing urban populations. Other animals – or parts of them – were not only processed by but did become integral components of the new industrial machinery in a quite literal sense. Tanned hides from bison killed en masse in the North American grasslands combined durability and flexibility to make excellent heavy leather belts needed to transmit power and motion in industrial machinery, while the spermaceti substance found in the head cavity of sperm whales was valued as a non-corrosive lubricant reducing friction and wear in moving parts. The hunting of both species ramped up significantly as industrial demand increased in and beyond North America and Europe, exemplifying the ways in which industrialised economies increasingly linked distant regions and ecosystems through socioecological "teleconnections."[30] The burgeoning international trade in wild animals, many of whom were captured in colonial locales, was another element in the production of these distal causal linkages concomitant to the globalisation of capital. By the late 1800s, American zoos were increasingly interested in acquiring animals from all over the world, and so were enterprising American showmen and those involved in the incipient trade in "exotic" pets, as both circus menageries and pet dealerships worked to monetise the "encounter value" of living creatures by displaying or selling them to American audiences or customers.[31]

THIS COLLECTION

It is hard to overstate the extent to which assembling a collection of primary sources today benefits from the wealth of digitised source material stored online in repositories like the Internet Archive, HathiTrust, or the Biodiversity Heritage Library. The flipside, however, is that navigating and sifting through this enormous amount of material is challenging and time-consuming, especially because the research potentials that come with it are still hampered by relatively limited search capabilities that ultimately require lots of manual searching. More than that, even more recent digital archives often remain distinctly anthropocentric in the ways in which source materials are described and catalogued, which means that looking for the proverbial "animal trace" can quickly devolve into a treasure hunt without a map. In addition to the aforementioned platforms, I have drawn upon a variety of other open access or subscription databases, including the Library of Congress's Chronicling America project, collections of historical newspapers and periodicals by ProQuest, Readex, and the American Antiquarian Society, as well as collections hosted by individual universities and the archives of various individual publications. It probably does not require explanation that even the substantial amount of space granted by five volumes does not authorise any claim to comprehensiveness, and even less so in a collection like this one, which combines an extensive time frame with a broad thematic scope.

The relative prominence or obscurity of source material has not been a decisive factor in compiling the present collection, though I have generally opted for lesser-known authors and/or texts if they met my expectations regarding the points I wanted them to convey within their thematic context. Obviously, this collection is no exception to the general fact that any process of source selection is not only shaped by subjective assessment and preference but inevitably involves numerous exclusions. In conjunction with other editorial decisions – even seemingly innocuous ones like the relative placement of sources in their respective sections or volumes – selection influences how readers encounter a given subject matter, potentially guiding them toward particular conclusions or understandings, or highlighting specific aspects while downplaying others. None of the sources collected here are obligatory choices, and one can easily imagine another editor operating with the same basic structure of volumes and sections assembling a collection comprised of strikingly different material. There are also some clear omissions that may be more or less defensible depending on one's opinion on the purpose of a project like this one. The present collection is certainly "mammalocentric" (and most definitely "vertebratocentric"), reflecting the editor's decision to highlight prevailing cultural emphases, interests, and tendencies of the time instead of pushing into more "taxonomically inclusive" animal historical territory, which would of course have been a perfectly legitimate and worthwhile approach. While mammals – especially domesticates but also prominent and/or "charismatic" wild creatures – dominate the archival record, testifying to their central roles in social and economic life and the cultural imagination, this centrality of human-attributed significance in part reflects hierarchies of attention and visibility that may have comparatively little to say about such aspects as actual ecological relevance or biological prevalence. Moreover, omissions abound even from an avowedly mammalocentric perspective, in part due to a preference for depth over breadth. For example, instead of including several sources on dogs in pet cultural and urban contexts with the aim of providing insight into a spectrum of associated perspectives and relations, another editor might have decided to use the space so allocated for material on other creatures with the aim of highlighting the capaciousness of nineteenth-century multispecies landscapes. I encourage readers to be mindful of such trade-offs and editorial decisions, none of which are beyond critique.

Each of the five volumes that make up the present collection includes an introduction and editorial headnotes that operate on two levels of contextualisation, the former drawing a wider contextual circle by staking out the terrain covered by the volume as a whole, the latter providing more specific biographical and contextual information about author(s) and source(s). This division of tasks is also reflected in the way endnote references for further reading are included, with those cited in the introductions generally pointing readers to literature that offers a broader discussion of the various themes and topics covered in the volume and those cited in the headnotes being more specifically geared towards the issues mentioned in or surrounding the respective source material. Each volume also contains (mostly) brief editorial notes on concepts and terminology, persons, places, events, literary characters, or uncredited quotations mentioned or included in the sources but not central enough to be addressed directly in the respective headnotes. Some documents have been abbreviated not only for general reasons of space but also to eliminate redundancies and/or to maintain thematic focus and narrative flow. My desire to have the tables of contents spell out with as much precision as possible what the included material is about means that I have taken some freedoms with the titles for individual items, even though I have generally tried to stick to the original titles of the source materials reproduced here. For example, while I have used the original title for the extracts from *Incidents of a Whaling Voyage* in Vol. 5 because, in my estimation, the book's title reflects the content of the chosen extracts adequately enough, in other instances I have made modifications (as with the extracts from the *Harper's* article "My Life as a Slave" in Vol. 3, which has been adapted to "My Life as a Slave and Horseman") or assigned new titles (as with the *McClure's* article "The Psychology of Wild Animals" in Vol. 1, which has been adapted to "Animal Crime and Criminal Animals"). Original bibliographical information is, of course, provided with each source text.

Volumes I and II of this collection trace the significance of animals, and the concept or "problem" of animality more generally, in the currents of social and scientific thought during a period that saw a rapid expansion of print culture both in the U.S. and internationally. Volume I examines how evolving ideas about animals and animality – especially about animal minds – interacted with traditional concepts of human nature and could be employed to both defend and challenge claims to human uniqueness based on rigid distinctions between human and animal lives. Volume II extends this discussion to address the ways in which the nexus of ideas and claims surrounding the human-animal distinction became intertwined with human intergroup hierarchies and relations of power, including through what Claire Jean Kim has described as the synergistic dynamics between race and species as fundamentally co-implicating "taxonomies of power" that have shaped white settler society's relations with both its internal and external racialised others.[32] A second section traces the evolution of animal advocacy from early individual voices during the early republic to an organised movement after the Civil War, documenting how this new form of postbellum advocacy reflected a shift – however limited by structural constraints – from largely anthropocentric concerns with the social consequences of (allowing or enabling) cruelty towards other creatures to a broader moral consideration for nonhuman animals in their own right.

The structure of volumes III to V reflects the conventional distinction between domesticated and wild animals. This structure will likely make "intuitive" sense to the majority of the readers of the present collection, reflecting the ways in which the wild-domestic divide has played a formative role in structuring how members of European(-derived) societies have perceived, evaluated, and related to the human and nonhuman worlds around them. It should be clear, however, that this distinction is by no means a self-evident "natural" (biological and/or ecological) fact but has functioned as a powerful ideological construct not only through its imposition of an anthropocentric categorical scheme onto the complexities of nonhuman life but also because the "wild" and the "domestic" have appeared as intimate allies of another antithetical pair in the currents of settler colonial discourse, that of "savagery" and "civilisation," and thus been implicated in the legitimisation of violence against Indigenous peoples, animals, and environments throughout the long nineteenth century. Does reproducing this problematic binary in a collection like this one reinforce its cultural power? Such a concern is not without legitimacy, but my own perspective on this issue is that there is critical value in foregrounding dominant frameworks in order to tackle them in both their historical contingency and ongoing influence. The introductions of Volumes III and V as well as relevant headnotes in both volumes elaborate on how the wild-domestic divide informed contemporary cultural practices and social hierarchies – human and more-than-human – as well as environmental perceptions and interactions.

Volume III begins with an exploration of the role of domesticated animals in what for most of the nineteenth century remained a predominantly agrarian world (albeit one undergoing significant change) centred on the labour and products that could be extracted from animal bodies, covering the "education" of working animals, contemporary attempts at "improving" animal breeds to increase agricultural productivity and market value, the importance of veterinary medicine, and other topics. The second section demonstrates that, the "transportation revolution" of the antebellum decades notwithstanding, animals remained essential for moving around goods and people in both rural areas and the growing American cityscapes, where horses powered everything from carriages to streetcars, as well as in other nineteenth-century contexts that relied on equine, bovine, and, sometimes, canine traction. A final section provides insights into the roles of different types of domesticates in entertainment and competitive settings, ranging from the prestigious if embattled sport of Thoroughbred horse racing to pigeon fancying to the brutal spectacles of dog- and cockfighting, which quickly became the target of a spirited post-Civil War anti-cruelty movement. Volume IV then moves on to the topic of pet culture and its evolution as a cultural institution over the course of the long nineteenth century. In part driven by the antebellum rise of the American middle-class family and the sentimentalisation of (certain) human-animal relationships, by the turn of the twentieth century petkeeping had spawned an expansive industry that offered everything from gourmet pet foods and fashionable accessories to healthcare and boarding services. The rise of pet culture also had a significant impact on urban life, the topic of the volume's second section. Besides walking, sitting, or lying on sidewalks and being sold in stores and on street corners, in case of abandonment the animals swelled an ever-increasing population of canine and feline strays, which together with horses, pigs, cows, chicken, and urban wildlife were an integral part of how human denizens experienced life in the nineteenth-century city.

Lastly, the materials that make up Volume V cover three key areas of interaction and concern that shaped Americans' relations with wild animals. The sources in section one focus on hunting, a practice that among early republicans was still associated with the "savage" existence of Indian peoples and regarded as incompatible with the agrarian virtues they deemed essential but that eventually became emblematic of settler identity and masculinity and tangled up in the politics of race and class, especially as the practice became increasingly steeped in regulations and restrictions in the context of largely elite-driven conservation efforts. The second section examines practices and sites of animal display – specifically natural history museums, zoological gardens, and circus menageries – for commercial or educational purposes, highlighting how such displays evolved from the private collections and traveling exhibitions of the early republican and antebellum decades into significant institutions that shaped American perceptions of wild animals. The third and final section discusses the growing awareness of anthropogenic species extinction in U.S. society, focusing especially on the dramatic decline of the American bison and the passenger pigeon and the cultural and political responses to these losses, tracing a long-nineteenth-century arc that began with opposition to the very idea of extinction and concluded with Progressive-Era campaigns that managed to save the bison from the brink but amounted to too little too late for the pigeon.

There is not a single point in human history that has not been shaped by humans' polymorphous entanglements with other species. To slightly rephrase anthropologist Anna Tsing: human history is an interspecies relationship.[33] As we grapple with questions of environmental ethics and multispecies justice among the ravages of global capitalism, a critical look at the long nineteenth century reveals uncomfortable trajectories whose consequences are tightly woven into the structural realities of the present in and beyond North American and European contexts. Perhaps this collection may serve as a small contribution to the important task of illuminating how historical patterns of human-animal relations have shaped contemporary societal frameworks and ecological challenges.

In conclusion, I would like to express my sincere thanks to the staff at Routledge, especially Simon Alexander, for their immense patience with a project that accompanied its editor through a series of rough patches that significantly delayed its completion.

Notes

1 Jürgen Osterhammel, *The Transformation of the World: A Global History of the Nineteenth Century*, trans. Patrick Camiller (Princeton: Princeton University Press, 2014).
2 Jürgen Osterhammel, *The Transformation of the World: A Global History of the Nineteenth Century*, trans. Patrick Camiller (Princeton: Princeton University Press, 2014), 47, original emphases.
3 Now widely used, the idea of a long nineteenth century became popular with the publication of British historian Eric Hobsbawm's trilogy on the "ages" of revolution, capital, and empire (1962–87), which discusses major social, political, and economic transformations in Europe from the French Revolution in 1789 to the global conflagration of World War I in 1914. For a critical view on the use of a long-nineteenth-century

perspective in world history, see Peter N. Stearns, "Rethinking the Long 19th Century in World History: Assessments and Alternatives," *World History Connected* 9, no. 3 (2012), https://worldhistoryconnected.press.uillinois.edu/9.3/forum_stearns.html.

4 Matthew Edgeworth et al., "The Anthropocene Is More Than a Time Interval," *Earth's Future* 12, no. 7 (2024): e2024EF004831; Philip Gibbard et al., "The Anthropocene as an Event, Not an Epoch," *Journal of Quaternary Science* 37, no. 3 (2022): 395–99.

5 Charles Sellers, *The Market Revolution: Jacksonian America, 1815–1846* (New York: Oxford University Press, 1991); and more recently John Lauritz Larson, *The Market Revolution in America: Liberty, Ambition, and the Eclipse of the Common Good* (Cambridge: Cambridge University Press, 2009).

6 The term "fossil capital" is from Andreas Malm, *Fossil Capital: The Rise of Steam-Power and the Roots of Global Warming* (London: Verso, 2016). On industrialisation and urbanisation, see Robert C. Allen, *The British Industrial Revolution in Global Perspective* (Cambridge: Cambridge University Press, 2009); James C. Williams, "The American Industrial Revolution," in *A Companion to American Technology*, ed. Carroll W. Pursell (Malden: Blackwell, 2005), 31–51; Carroll W. Pursell, *The Machine in America: A Social History of Technology*, 2nd ed. (Baltimore: Johns Hopkins University Press, 2007); Richard L. Hills, *Power from Steam: A History of the Stationary Steam Engine* (Cambridge: Cambridge University Press, 1993); Jeremy Atack, Robert A. Margo, and Paul W. Rhode, "Industrialization and Urbanization in Nineteenth Century America," *Regional Science and Urban Economics* 94 (2022): 103678; Eric H. Monkkonen, *America Becomes Urban: The Development of U.S. Cities and Towns, 1780–1980* (Berkeley: University of California Press, 1988); Lisa Krissoff Boehm and Steven H. Corey, *America's Urban History*, 2nd ed. (New York: Routledge, 2024); Paul S. Boyer, *Urban Masses and Moral Order in America, 1820–1920* (Cambridge: Harvard University Press, 1992); Dorceta E. Taylor, *The Environment and the People in American Cities, 1600–1900s: Disorder, Inequality, and Social Change* (Durham: Duke University Press, 2009). On energy transitions, see David E. Nye, *Consuming Power: A Social History of American Energies* (Cambridge: MIT Press, 1998); Christopher F. Jones, *Routes of Power: Energy and Modern America* (Cambridge: Harvard University Press, 2014); Charles Blanchard, *The Extraction State: A History of Natural Gas in America* (Pittsburgh: University of Pittsburgh Press, 2021).

7 Edwin T. Layton, "Mirror-Image Twins: The Communities of Science and Technology in 19th-Century America," *Technology and Culture* 12, no. 4 (1971): 562; Edwin T. Layton, "American Ideologies of Science and Engineering," *Technology and Culture* 17, no. 4 (1976): 688. On the distinction between technological knowledge and scientific knowledge, see John M. Staudenmaier, *Technology's Storytellers: Reweaving the Human Fabric* (Cambridge: MIT Press, 1984).

8 Paul B. Israel, "Inventing Industrial Research: Thomas Edison and the Menlo Park Laboratory," *Endeavour* 26, no. 2 (2002): 48–54; Leonard S. Reich, *The Making of American Industrial Research: Science and Business at GE and Bell, 1876–1926* (Cambridge: Cambridge University Press, 1985).

9 For discussions of American science in the long nineteenth century, see, for example, George H. Daniels, *American Science in the Age of Jackson* (Tuscaloosa: University of Alabama Press, 1968); Walter H. Conser, *God and the Natural World: Religion and Science in Antebellum America* (Columbia: University of South Carolina Press, 1993); Charles E. Rosenberg, *No Other Gods: On Science and American Social Thought*, 2nd ed. (Baltimore: Johns Hopkins University Press, 1997); Philip J. Pauly, *Biologists and the Promise of American Life: From Meriwether Lewis to Alfred Kinsey* (Princeton: Princeton University Press, 2000); and the essays in Georgina M. Montgomery and Mark A. Largent, eds., *A Companion to the History of American Science* (Chichester: John Wiley & Sons, 2016); and Timothy W. Kneeland, ed., *The Routledge History of American Science* (New York: Routledge, 2022).

10 For comprehensive discussions of American expansionism, see Walter Nugent, *Habits of Empire: A History of American Expansion* (New York: Knopf, 2008); Walter L. Hixson, *American Settler Colonialism: A History* (New York: Palgrave Macmillan, 2013); Adam Burns, *American Imperialism: The Territorial Expansion of the United States, 1783–2013* (Edinburgh: Edinburgh University Press, 2017); Paul Frymer, *Building an American Empire: The Era of Territorial and Political Expansion* (Princeton: Princeton University Press, 2017); Daniel Immerwahr, *How to Hide an Empire: A Short History of the Greater United States* (New York: Farrar, Straus and Giroux, 2019). On slavery and its interplay with territorial expansion, see Walter Johnson, *River of Dark Dreams: Slavery and Empire in the Cotton Kingdom* (Cambridge: Harvard University Press, 2013); David Brion Davis, *The Problem of Slavery in the Age of Emancipation* (New York: Alfred A. Knopf, 2014); Adam Rothman, "Slavery and National Expansion in the United States," *OAH Magazine of History* 23, no. 2 (2009): 23–29; William W. Freehling, *The Road to Disunion: Secessionists Triumphant, 1854–1861* (New York: Oxford University Press, 2007). Though sometimes overstated to the point of technological determinism in its importance as a driving force of antebellum slavery, the cotton gin – patented by Eli Whitney in 1794 – certainly made cotton plantations operated with enslaved labour immensely profitable enterprises. On the global cotton boom, see Sven Beckert, *Empire of Cotton: A Global History* (New York: Knopf, 2014).

11 James Walvin, "The Slave Trade, Quakers, and the Early Days of British Abolition," in *Quakers and Abolition*, ed. Brycchan Carey and Geoffrey Plank (Urbana: University of Illinois Press, 2014), 165–79; Seymour Drescher, *Abolition: A History of Slavery and Antislavery* (Cambridge: Cambridge University Press, 2009); John R. Oldfield, *Transatlantic Abolitionism in the Age of Revolution: An International History of Anti-Slavery, c.1787–1820* (Cambridge: Cambridge University Press, 2013); John R. Oldfield, *The Ties That Bind: Transatlantic Abolitionism in the Age of Reform, c. 1820–1866* (Liverpool: Liverpool University Press, 2020). On antislavery and women's rights, see Martha S. Jones, *All Bound up Together: The Woman Question in African American Public Culture, 1830–1900* (Chapel Hill: University of North Carolina Press, 2007); as well as the essays in Kathryn Kish Sklar and James Stewart, eds., *Women's Rights and Transatlantic Antislavery in the Era of Emancipation* (New Haven: Yale University Press, 2007); and Elizabeth J. Clapp and Julie Roy Jeffrey, eds., *Women, Dissent and Anti-Slavery in Britain and America, 1790–1865* (Oxford: Oxford University Press, 2011). For a broader discussion of the rise of modern social movements, see Charles Tilly, Ernesto Castañeda, and Lesley J Wood, *Social Movements, 1768–2018*, 4th ed. (New York: Routledge, 2020).

12 John Quincy Adams, "Society and Civilization," *The American Review* 2, no. 1 (July 1845): 80–89, here 80, original emphasis. Also see Dominik Ohrem, "A Declaration of Interdependence: American History and the Challenges of Postanthropocentric Historiography," in *American Beasts: Perspectives on Animals and Animality in U.S. Culture, 1776–1920*, ed. Dominik Ohrem (Berlin: Neofelis, 2017), 9–48.

13 Mieke Roscher, André Krebber, and Brett Mizelle, "Writing History After the Animal Turn: An Introduction to Historical Animal Studies," in *Handbook of Historical Animal Studies*, ed. Mieke Roscher, André Krebber, and Brett Mizelle (Berlin: De Gruyter, 2021), 1–18, here 12; Gary Wilder, "From Optic to Topic: The Foreclosure Effect of Historiographic Turns," *The American Historical Review* 117, no. 3 (2012): 723–45. For examples of scholarship on the historical dynamics between race and animality, see Kay Anderson, *Race and the Crisis of Humanism* (London: Routledge, 2007); Claire Jean Kim, *Dangerous Crossings: Race, Species, and Nature in a Multicultural Age* (New York: Cambridge University Press, 2015); Lindgren Johnson, *Race Matters, Animal Matters: Fugitive Humanism in African America, 1838–1934* (New York: Routledge, 2018); Bénédicte Boisseron, *Afro-Dog: Blackness and the Animal Question* (New York: Columbia University Press, 2018).

14 On "anthromes," see Erle C. Ellis, "Anthropogenic Transformation of the Terrestrial Biosphere," *Philosophical Transactions of the Royal Society A: Mathematical, Physical and Engineering Sciences* 369, no. 1938 (2011): 1010–35; Erle C. Ellis, Arthur H. W. Beusen, and Kees Klein Goldewijk, "Anthropogenic Biomes: 10,000 BCE to 2015 CE," *Land* 9, no. 5 (2020): 129.

15 On the concept of social structure, see John Levi Martin, *Social Structures* (Princeton: Princeton University Press, 2009); Charles Crothers, *Social Structure* (Abingdon: Routledge, 1996); Charles Crothers, "Social Structure: History of the Concept," in *International Encyclopedia of the Social & Behavioral Sciences*, ed. James D. Wright, 2nd ed., vol. 22 (Amsterdam: Elsevier, 2015), 719–26.

16 William H. Sewell, "Historical Events as Transformations of Structures: Inventing Revolution at the Bastille," *Theory and Society* 25, no. 6 (1996): 841–81, here 842.

17 William H. Sewell, *Logics of History: Social Theory and Social Transformation* (Chicago: University of Chicago Press, 2005), 126.

18 William H. Sewell, *Logics of History: Social Theory and Social Transformation* (Chicago: University of Chicago Press, 2005), 146.

19 Erika Cudworth, *Social Lives with Other Animals: Tales of Sex, Death and Love* (Houndmills: Palgrave Macmillan, 2011), 176.

20 See, for example, Sarah E. McFarland and Ryan Hediger, eds., *Animals and Agency: An Interdisciplinary Exploration* (Leiden: Brill, 2009); Helen Steward, "Animal Agency," *Inquiry* 52, no. 3 (2009): 217–31; Hans-Johann Glock, "Animal Agency," in *A Companion to the Philosophy of Action*, ed. Timothy O'Connor and Constantine Sandis (Chichester: Wiley-Blackwell, 2010), 384–92; Jennifer A. Martin, "When Sharks (Don't) Attack: Wild Animal Agency in Historical Narratives," *Environmental History* 16, no. 3 (2011): 451–55; Bob Carter and Nickie Charles, "Animals, Agency and Resistance," *Journal for the Theory of Social Behaviour* 43, no. 3 (2013): 322–40; David Gary Shaw, "The Torturer's Horse: Agency and Animals in History," *History and Theory* 52, no. 4 (2013): 146–67; Chris Pearson, "History and Animal Agencies," in *The Oxford Handbook of Animal Studies*, ed. Linda Kalof (New York: Oxford University Press, 2017), 240–57; Amanda Rees, "Animal Agents: Historiography, Theory and the History of Science in the Anthropocene," *BJHS Themes* 2 (2017): 1–10; Philip Howell, "Animals, Agency, and History," in *The Routledge Companion to Animal-Human History*, ed. Hilda Kean and Philip Howell (Abingdon: Routledge, 2018).

21 William H. Sewell, *Logics of History: Social Theory and Social Transformation* (Chicago: University of Chicago Press, 2005), 251, 245.

22 Bob Carter and Nickie Charles, "Animals, Agency and Resistance," *Journal for the Theory of Social Behaviour* 43, no. 3 (2013): 322–40, 331.

23 John Berger, "Why Look at Animals?," in *About Looking*, by John Berger (1980; repr., New York: Vintage Books, 1991), 3–28, here 9.

24 John Berger, "Why Look at Animals?," in *About Looking*, by John Berger (1980; repr., New York: Vintage Books, 1991), 3–28, 9, 5, 11.

25 Jonathan Burt, "John Berger's 'Why Look at Animals?': A Close Reading," *Worldviews: Global Religions, Culture, and Ecology* 9, no. 2 (2005): 203–18; Anat Pick, "Why Not Look at Animals?," *NECSUS: European Journal of Media Studies*, 2015, www.necsus-ejms.org/why-not-look-at-animals/.

26 Edmund Russell, "Introduction: The Garden in the Machine: Toward an Evolutionary History of Technology," in *Industrializing Organisms: Introducing Evolutionary History*, ed. Philip Scranton and Susan R. Schrepfer (New York: Routledge, 2004), 1–16, here 2.

27 Ann Norton Greene, *Horses at Work: Harnessing Power in Industrial America* (Cambridge: Harvard University Press, 2008); Gervase Phillips, "Writing Horses into American Civil War History," *War in History* 20, no. 2 (2013): 160–81; Margaret Evans, "Pit Ponies: Ghosts of the Coal Mines," July 7, 2023, www.horsejournals.com/popular/

history-heritage/pit-ponies-ghosts-coal-mines; also see Les Mitchell, "Stolen Children of the Endless Night: A Critical Account of the Lives of British Pit Ponies," in *Animals as Experiencing Entities: Theories and Historical Narratives*, ed. Michael J. Glover and Les Mitchell (Cham: Springer, 2024), 259–93.

28 The term "commodi-faunation" is from Tony Weis, "Ghosts and Things: Agriculture and Animal Life," *Global Environmental Politics* 18, no. 2 (2018): 134–42.

29 Dinesh Wadiwel, *The War Against Animals* (Boston: Brill, 2015), 65.

30 John R. McNeill, "Bison, Elephants, and Sperm Whales: Keystone Species in the Industrial Revolution," *Historical Social Research* 48, no. 1 (2023): 146–69; Jennifer Hansen, "A Tanner's View of the Bison Hunt: Global Tanning and Industrial Leather," in *Bison and People on the North American Great Plains: A Deep Environmental History*, ed. Geoff Cunfer (College Station: Texas A&M University Press, 2016), 227–44. Also see John R. McNeill, "Cheap Energy and Ecological Teleconnections of the Industrial Revolution, 1780–1920," *Environmental History* 24, no. 3 (2019): 492–503; Cecilie Friis et al., "From Teleconnection to Telecoupling: Taking Stock of an Emerging Framework in Land System Science," *Journal of Land Use Science* 11, no. 2 (2016): 131–53.

31 Maan Barua, "Lively Commodities and Encounter Value," *Environment and Planning D: Society and Space* 34, no. 4 (2016): 725–44; Rosemary-Claire Collard, *Animal Traffic: Lively Capital in the Global Exotic Pet Trade* (Durham: Duke University Press, 2020).

32 Claire Jean Kim, *Dangerous Crossings: Race, Species, and Nature in a Multicultural Age* (New York: Cambridge University Press, 2015), esp. chapter 2.

33 "*Human nature is an interspecies relationship.*" Anna Tsing, "Unruly Edges: Mushrooms as Companion Species," *Environmental Humanities* 1 (2012): 141–54, here 144, original emphasis.

VOLUME I INTRODUCTION

The sources in the first two volumes of this collection provide insights into the varied influence of animals and animality on American thought during the long nineteenth century. In the late 1700s, proto-anthropological questions about human origins, human nature, and the boundary between humans and (all) other animals began to assume a central position among a cohort of European and American intellectuals. There existed, of course, a long tradition of thought about the characteristics of human beings and their (privileged) place on earth and in the cosmos before the turn of the nineteenth century. But as Michel Foucault has famously argued in *The Order of Things*, what first developed during this period was an "epistemological consciousness of man as such" and of a "specific domain proper to man" that would be reflected in the transition from natural history to the "sciences of man" – including what would develop into the disciplines of anthropology, archaeology, and linguistics – as part of a broader, diversified scientific landscape populated by specialised professionals working for a growing number of scientific institutions.[1]

A few words are in order about the meaning and scope of the term "American thought." First, the qualifier "American" should be understood in terms of emphasis rather than exclusivity. Beginning with the exchange and circulation of ideas through the correspondence networks of a transatlantic Republic of Letters that predated the national period, the currents of American thought – scientific, philosophical, political, religious, or otherwise – were never truly isolated from the rest of the world, much less so from Europe.[2] At the same time, much of the "Americanness" of American thought during the long nineteenth century can be traced not only to attempts at complementing the republic's nominal political independence with a sense of national identity emancipated from its Old World ties but also, and more importantly, to the social, political, and economic structure of a white settler society whose growth had been premised on the elimination of Native peoples and the exploitation of enslaved labour. As will be discussed in more detail in Volume II, nowhere was this more evident than in the ways in which discussions around and definitions of "the human" readily lent themselves to an implicit or explicit defence of racial hierarchy and oppression. The somewhat passive and innocent ring of the term "thought" should thus not distract us from

DOI: 10.4324/9781003032717-1

its imbrication with processes of knowledge production that themselves reflected and reinforced contemporary sociomaterial relations of power both within U.S. society and on the continent more broadly.

A second and somewhat related point I wish to address pertains to the type(s) of thought exemplified by the sources collected here. Taken together, most of the material in the first and second volumes exemplifies the strong intersections between American science and American social thought. Discussing this dynamic relationship in a classic book on the subject, Charles E. Rosenberg identifies as its most important element "the increasing emotional relevance of science, its expanding role as an absolute, able to justify and even motivate the behavior of particular individuals" as well as the "vocabulary" and "supply of images" it has lent to social thought in a way that "variously suggested, explained, justified, even helped dictate social categories and values."[3] This relationship was by no means unidirectional but was defined by a degree of reciprocity in that American society itself articulated "demands upon the scientist's work and thought" and shaped scientific work to a degree that affected even "the internal texture of scientific ideas."[4]

The expansion of print culture in the first decades of the nineteenth century was crucial to both the beginning institutionalisation and the early popularisation of American science. According to Heather Haveman, it was largely due to "the development of material and cultural supports for magazine publishing" such as steam-powered presses and the growth of the postal network that the magazine founding rate eventually outstripped the failure rate, with the number of magazines in print growing from just above 80 to almost 350 between 1810 and 1830 (and 1000 by 1860) and their median life span increased times five.[5] Even before the emergence of popular science magazines like *Scientific American* and *Popular Science Monthly* (founded in 1845 and 1872, respectively), articles printed in various general interest magazines – many of them short-lived, others, like Boston's *North American Review* (founded in 1815) more enduring – exposed an engaged nonspecialist reading public to specialist debates on which it asserted a degree of influence rather than merely assuming a role of passive recipiency.[6]

The sources collected in this first volume are centred on long-nineteenth-century negotiations of human-animal kinship and/versus notions of human uniqueness. They exemplify, weigh in on, or critique various contemporary attempts at defining human specificity in terms of what philosophers have come to refer to as "anthropological difference." While all animal species are "specific" in some way, the idea of (an) anthropological difference moves well beyond this innocuous observation in that it aims to establish, in Hans-Johann Glock's words, "a difference with a difference" by focusing on

> features of homo sapiens that (a) set us apart "categorially" or "essentially" from all other animals; (b) are fundamental, in that (all) other relevant differences derive from them; (c) are important, notably to our self-image, for instance because they assure us of a higher spiritual or moral status than animals.[7]

Whatever criterion or set of criteria they were based on, such generalised distinctions between human and animal tended to lump together the multiformity of nonhuman animal life under a single label in an act of epistemic violence, as Jacques Derrida has famously noted[8] – even if the fabrication of this amorphous (non-)entity called "the animal" in fact often relied on discussions of very specific (types of) animals.

ANIMALS, ANIMALITY, AND ANTHROPOLOGICAL DIFFERENCE

In an 1872 review of French popular science writer Louis Figuier's book *The Human Race*, pioneering American anthropologist Lewis Henry Morgan argued that "the special creation of man" was "the question of questions in modern science."[9] Formulated over a decade after the publication of Charles Darwin's *On the Origin of Species* (1859) and shortly after that of *The Descent of Man* (1871) – texts which amounted to what Carrie Rohman calls a "radical philosophical blow to anthropocentrism" – Morgan's assessment testifies to the ongoing contestation of traditional concepts of the human through the ideas of common descent and evolutionary continuity with nonhuman creatures.[10] As Carl Degler explains, Darwin's work was "the culmination rather than the initiation of a line of thought that saw evolutionary change in man and nature," given that Charles' grandfather Erasmus Darwin and philosopher Denis Diderot (among others) had already speculated about the possibility of organic evolution before Jean-Baptiste Lamarck proposed a more systematic formulation of the idea in his *Philosophie Zoologique* (1809).[11] The emergence of evolutionary thought meant that Euro-Americans had to come to terms with the ever more forceful suggestion that there were, in fact, no humans and animals but only humans and other animals, ushering in what Raymond Corbey refers to as a "cosmological sea-change" that resulted in an anxious awareness of the uncertain position of humans within the natural hierarchical order commonly understood in terms of a fixed Chain of Being.[12] Because this emergent cosmology "no longer explained humans *meta*physically . . . but traced their origin to the physics of lowly animals," defenders of human uniqueness were forced "into reluctant retreats from, and renegotiations of, the notion of [their] own [human] specialty," resulting in hard-fought battles about exactly how much of dignified humanity had to be conceded to the sphere of animality.[13]

While evolutionary theory's belief in a transformative dynamism operative in the natural world was at odds with the more static models that dominated eighteenth-century natural history, the works of naturalists like Carl Linnaeus and Georges-Louis Leclerc de Buffon articulated their own challenges to established ideas about human uniqueness. What they implied in a general sense was the troubling possibility of reinterpreting the human being as "a purely zoological object, subjected to all the fragmenting powers of history and geography operative on any other species," and it was this interpretive possibility, once introduced into the currents of social and scientific thought, that could not be easily undone.[14]

Discussing the emergence of what he terms the "natural historical gaze," Phillip Sloan argues that the eighteenth century inaugurated "a human science that institutionally, as well as conceptually, focused upon a 'zoological history' of the human species, now abandoned to the vicissitudes of time and space," and, eventually, to the "implications of ethnography, historical geology, animal biogeography, and cultural history."[15] Moreover, this proto-anthropological purview of much of natural history was confronted with the expanding horizons of the human concomitant to the expansion of geographical horizons through exploratory and colonial projects in different parts of the world, a process that opened up a complex "human space" and "rais[ed] the prospects of a plurality of human species."[16]

In the decades leading up to Darwin's *Origin*, the interest in Morgan's "question of questions" had already begun to reverberate beyond the confines of scientific inquiry more narrowly conceived, diffusing into the spheres of public discourse through a debate that spanned the domains of natural history, philosophy, theology, and other specialist and nonspecialist fields (or, perhaps more to the point, what would later be accepted as the specialist expertise of the sciences was yet to be neatly distinguished from nonspecialist modes of knowledge production). The involvement of a nonspecialist but interested reading public was fostered by a rapid expansion of print culture – the period between 1825 and 1850 was already recognised among contemporaries as the (first) "golden age of periodicals"[17] – and the growing number of periodicals, short-lived as many of them were, functioned less as unidirectional avenues for the cultural reception of scientific debates but actively shaped these debates themselves.

The broad interest in the problem of human nature went well beyond the satisfaction of philosophical or scientific curiosity.[18] In an 1851 treatise on *The Philosophy of Human Nature*, Francis E. Brewster points to the "undeniable omission of scientific research" regarding the "occult mysteries of human nature, a thorough knowledge of which is so intimately essential to man's social safety and moral elevation."[19] For New York physician and Harvard graduate Martyn Paine, the importance of this subject, particularly in light of the "revolutionary spirit" that sought to reframe the human being in "materialist" terms in conflict with religious doctrine (a "terrific crisis"), was "not confined to our own [medical] science, nor to general literature and philosophy, but strikes at the more absolute foundations of society."[20] And in an 1873 article in the *Princeton Review*, Scottish philosopher and minister Henry Calderwood refers to "[t]he study of man" as "a pressing demand in our times," emphasising the need for "a full and systematized knowledge of human nature."

> We are in some things at least like the animals in our actions. How far are we like them, and in what ways? We are in other respects unlike the animals in our conduct. How far does ascertained dissimilarity separate us from them? Just behind these questions may rise the other, How far is the hypothesis feasible that human nature is only a development of lower orders of organized being?[21]

An 1845 article in the *Knickerbocker* defines the study of human nature as an inquiry into "all that essentially belongs to man, physically, intellectually and morally, and whether to him exclusively, or in common with other beings."[22]

For the most part, however, the search for human nature was less motivated by a search for commonalities but a quest for anthropological difference, an effort to "determine human specificity over and against those beings who/that threaten to undermine that specificity."[23] The search for this distinctively human essence has been characterised by what anthropologist Tim Ingold has called the "logic of inversion," inverting the question "What makes humans animals of a particular kind?" to "What makes humans different in kind from animals?"[24] One way of thinking about anthropological difference is as a powerful a priori, or even an ethical imperative, that implicitly pre-shapes all inquiries into human being(s), human origins, and human-animal kinship. That is, rather than representing the resolution of an open-minded inquiry, anthropological difference functions as the hidden premise of an investigation whose outcome is, to a significant extent, always already a foregone conclusion.

An insightful example of contemporary attempts at delineating anthropological difference is provided by an 1845 article titled "Society and Civilization" by former (sixth) president John Quincy Adams. In it, Adams echoes a widespread belief in claiming that, while it was "the property of all animated nature to be *gregarious*," humans alone could be considered properly *social* beings in a sense that truly warranted the use of this concept (for a counterperspective, see C. A. Ellwood, this volume).[25] Adams writes:

> The beasts of the forest and the fields are drawn by the laws of their nature together in *herds* – the birds of the air in *flocks* – the fishes of the ocean in *sculls*; reptiles *cluster* in the recesses of the earth, and the insect tribes no sooner quicken from their shells into life, than they people the sunbeam in *swarms*.

But the purpose of this propensity to congregate mostly remained a mystery because at least according to Adams it did not seem to have one.[26] Such forms of animal togetherness amounted to little more than a co-presence of animate bodies – conducive to neither "the subsistence of the individual" nor the "continuance of the species" and affording "no means of intellectual communion, where there is no mind to cultivate."[27] For Adams, what qualified as sociality was grounded in mutually reinforcing assumptions about both the mode of relationality that underpinned definitions of the social *and* about the types of beings that qualified for inclusion into this sphere of relationality. Society, that is, had to be understood in terms of rational intersubjectivity, and it was only in the interchanges that defined a collectivity of human minds that such a mode of relating could find its appropriate expression. It was through the interactions with other members of their species that humans were able to exercise their "prerogative of reason and faculty of speech" while, conversely, humans deprived of such relations would ultimately find themselves reduced to "a condition with no more indication of mind than there is in the Ourang Outang of Africa."[28]

Adams' distinction between human sociality and animal gregariousness ties in with two more prominent criteria wielded in nineteenth-century arguments for anthropological difference – reason and language. In line with a longer tradition that conceived of mind/soul and body as ontologically separate, the capacity for rational thought was attributed to what some referred to as an "immaterial principle" that was irreducible to the (human) animal body and articulated in rejection of the "materialist" tendencies that were criticised by commentators like Paine and attributed to fields like phrenology, which tied the mind back to the body with its claim that mental capacities were rooted in a range of cerebral "organs" and thus within the physical structure of the brain (see "Phrenology and Animal 'Character' in the Antebellum Period," this volume). The description of this mysterious immaterial principle by Harvard zoologist Louis Agassiz and conchologist Augustus Addison Gould in their 1848 *Principles of Zoölogy* affords some insights into its rather paradoxical constitution.

> Besides the material substance of which the body is constructed, there is also an immaterial principle, which, though it eludes detection, is none the less real, and to which we are constantly obliged to recur in considering the phenomena of life. It originates with the body, and is developed with it, while yet it is totally apart from it.[29]

The insistence on an immaterial principle allowed for a clear separation of humans from other forms of animal life in a way that remained largely unaffected or was even reinforced by the growing evidence for humans' evolutionary kinship with nonhuman creatures. It should be noted, however, that by the turn of the nineteenth century, an alternative form of exceptionalist discourse that Kay Anderson and Colin Perrin have termed "anatomical humanism" operated with the claim that humans' bipedal upright posture and gait were a unique *physical* differentiator from the rest of animalkind (for a perspective that combines physical and mental arguments for human uniqueness, see F. A. Rauch, this volume).[30] An 1852 article in the *American Review* explains that even in their "organic and sensuous attributes" – including "the upright figure, beauty of outline, universal grace of motion, smoothness and purity of skin, a countenance equally beautiful in motion or repose" – human beings were vastly superior to other animals to such an extent that "[f]rom the ape to the man there is no transition, but an abyss of differences."[31]

Rather than insisting on abyssal difference, others grappled with the strange familiarity between human and nonhuman primates, their uncanny "resemblance and identity of structure" and "horrible human likeness" (as a writer for *Putnam's Monthly* has it) that was supported by recent work in comparative anatomy.[32] Augustus R. Grote, in an 1878 paper on "Man and his Structural Affinities," describes the body of apes as "that of a human being, except that the arms are proportionally longer, and the legs shorter, than in the average man. The face is very human from the structure and position of the eyes and ears."[33] The hands of the gorilla – a creature of enormous fascination whose existence had only been confirmed recently – were "used for coarser work than man's" but "not for work very different in kind."[34] Beyond physical resemblances, contemporary observers like Arthur E. Brown or Richard L. Garner were interested in nonhuman primates' "powers of reasoning" and moral capacities, even if these were "not altogether such as would serve for the ordering of a human community," and in translating their "speech" (see A. E. Brown; R. L. Garner, this volume).

Among those whose case for human uniqueness relied on the capacity for rational thought, the latter was frequently contrasted with the notion of instinct, "that half-brother of reason," as Virginia planter and surveyor William Byrd puts it in reference to the behaviour of beavers – animals frequently evoked in nineteenth-century debates about animal mentation due to what some considered to be the obvious intelligence expressed in their architectural feats and future-oriented behaviour.[35] As "the determined action of vital forms without the agency of intellect," most people sharply distinguished instinct from reason, which denoted "[a]ction with the agency of intellect – a conclusion deduced from the relations of things – cause and effect."[36] However, not everyone agreed with such a strict distinction and/or the endeavour to bring this distinction into alignment with the human-animal boundary. The author of an 1857 article for *The Ladies' Repository* insists:

> The contrast between man and beast is *not a contrast between reason and instinct*, for it can be readily shown that these two mental qualities . . . are by no means inseparable; that in point of fact the dog is endowed with reason as well as instinct, and that man performs many actions which are purely instinctive, as well as those which are prompted by reason.[37]

Discussing the common usage of the term in *The American Beaver and His Works* (1868), Morgan criticises the concept of instinct itself, arguing that it was of dubious scientific value and that its main purpose consisted in defending the dignified position of the human mind as something distinguished in kind rather than degree from the "thinking principle" operative among animalkind more broadly (see L. H. Morgan, this volume). Two decades later, in his own discussion of instinct in relation to anthropological difference, philosopher-psychologist William James rejects the widespread assumption that, in stark contrast to (all) other animals, humans were defined by their deliberative rationality and at best minimally instinctual creatures (see W. James, this volume). Reason versus instinct remained a mainstay of investigations into the human-animal boundary in the closing decades of the century, which saw the advent of the field of comparative psychology and a novel emphasis on studying animal behaviour in controlled experimental settings (see E. L. Thorndike; M. F. Washburn, this volume).

The presence or absence of language in nonhuman species was a much-discussed question throughout the nineteenth century, and here the case for anthropological difference often relied on the assumption of an intimate connection between language and the capacity for rational thought.[38] Early American fiction writer William Cardell summarises this assumption in his *Essay on Language* (1825): "Every process of the reasoning powers, beyond the immediate perception of sensible objects, depends on the structure of speech," an absence of the latter indicating an absence of the former.[39] Thus, while animals were able to communicate via visual and auditory (as well as tactile and olfactory) modes of *expression*, language, many believed, did not emanate from the animal body – human or nonhuman – but was a unique product of the mind. In this way, as Susan Pearson explains, nineteenth-century debates about the existence of language in animals were not only bound up with the issue of defining the fundamental elements of language in a more general sense but also with "speculation about whether the boundaries of humanity and, more broadly, the moral community, were defined in terms of minds or bodies."[40] Animal advocates like Henry Bergh (see Volume Two) mobilised the absence of (human-like) language in animals to make a case for the moral imperative of speaking *for* this oppressed caste of "speechless slaves."[41]

Though widespread, the idea that language was limited to humans alone was not the only position. The most feasible way for those willing to extend the capacity for language to the nonhuman world was either thinking of animals as minded and thus potentially linguistic beings (retaining the assumed linkage between language and reason) or, alternatively, uncoupling language from mind and complicating distinctions between animal expression and human language, allowing for the possibility of conceptualizing language more inclusively in (inter)corporeal terms. Shifting the focus from animals' supposed lack of linguistic ability to humans' incapacity to understand animal languages, a writer for *Putnam's Monthly* argues that the problem was not that animals did not have language but that they communicated in "unknown tongues" (such as the "mysterious language of features") eluding human comprehension.[42] Lacking a "phonetic language," even the simplest of animals at least had "a language of instinct" to make "themselves understood by their own race and by their enemies."[43] For those who tried to make sense of the boundary between humans and animals in terms of linguistic ability, this boundary could thus be construed either as a kind of metaphysical abyss or in terms of a perhaps less insuperable chasm of mutual incomprehension that conceived of animals as more or less enigmatic others. "Were we to judge a strange man by his actions, his speech being wholly unintelligible to us," writer and poet Bayard Taylor suggests with regard to the quasi-agnostic latter perspective, "we should give him the credit of a positive conscience in like circumstances. Why should we withhold it from an animal?"[44]

Notes

1 Michel Foucault, *The Order of Things: An Archaeology of the Human Sciences*, trans. Alan Sheridan (New York: Pantheon Books, 1971), 336. On the (transatlantic) development of the "sciences of man," see Merle Curti, *Human Nature in American Thought: A History* (Madison: University of Wisconsin Press, 1980), chapter 6; John

C. Greene, *American Science in the Age of Jefferson* (Ames: Iowa State University Press, 1984), chapters 12–14; Christopher Fox, Roy Porter, and Robert Wokler, eds., *Inventing Human Science: Eighteenth-Century Domains* (Berkeley: University of California Press, 1995); Richard Olson, "The Human Sciences," in *The Cambridge History of Science, Vol. 4: Eighteenth-Century Science*, ed. Roy Porter (Cambridge: Cambridge University Press, 2003), 436–62; Christopher J. Berry, "The Rise of the Human Sciences," in *Scottish Philosophy in the Eighteenth Century, Volume I: Morals, Politics, Art, Religion*, ed. Aaron Garrett and James A. Harris (Oxford: Oxford University Press, 2015), 283–322. On American science in the long nineteenth century more generally, see, for example, George H. Daniels, *American Science in the Age of Jackson* (Tuscaloosa: University of Alabama Press, 1968); Walter H. Conser, *God and the Natural World: Religion and Science in Antebellum America* (Columbia: University of South Carolina Press, 1993); as well as the essays in Georgina M. Montgomery and Mark A. Largent, eds., *A Companion to the History of American Science* (Chichester: John Wiley & Sons, 2016); and Timothy W. Kneeland, ed., *The Routledge History of American Science* (New York: Routledge, 2022). On American natural history, see Volume 5.

2 See, for example, Susan Scott Parrish, *American Curiosity: Cultures of Natural History in the Colonial British Atlantic World* (Chapel Hill: University of North Carolina Press, 2006); Londa Schiebinger, "Scientific Exchange in the Eighteenth-Century Atlantic World," in *Soundings in Atlantic History: Latent Structures and Intellectual Currents, 1500–1830*, ed. Bernard Bailyn and Patricia L. Denault (Cambridge: Harvard University Press, 2009), 294–328; Mark A. Peterson, "*Theopolis Americana*: The City-State of Boston, the Republic of Letters, and the Protestant International, 1689–1739," in *Soundings in Atlantic History: Latent Structures and Intellectual Currents, 1500–1830*, ed. Bernard Bailyn and Patricia L. Denault (Cambridge: Harvard University Press, 2009), 328–70; Susan Manning and Francis D. Cogliano, eds., *The Atlantic Enlightenment* (Aldershot: Ashgate, 2008); William Max Nelson, "The Atlantic Enlightenment," in *The Atlantic World*, ed. D'Maris Coffman, A. B. Leonard, and William O'Reilly (London: Routledge, 2015), 650–66.

3 Charles E. Rosenberg, *No Other Gods: On Science and American Social Thought*, 2nd ed. (Baltimore: Johns Hopkins University Press, 1997), 1.

4 Charles E. Rosenberg, *No Other Gods: On Science and American Social Thought*, 2nd ed. (Baltimore: Johns Hopkins University Press, 1997), 2.

5 Heather A. Haveman, *Magazines and the Making of America: Modernization, Community, and Print Culture, 1741–1860* (Princeton: Princeton University Press, 2015), 105. On periodicals in nineteenth-century American society, see, for example, Simon Baatz, "'Squinting at Silliman': Scientific Periodicals in the Early American Republic, 1810–1833," *Isis* 82, no. 2 (1991): 223–44; Trish Loughran, *The Republic in Print: Print Culture in the Age of U.S. Nation Building, 1770–1870* (New York: Columbia University Press, 2007); Robert A. Gross and Mary Kelley, eds., *An Extensive Republic: Print, Culture, and Society in the New Nation, 1790–1840* (Chapel Hill: University of North Carolina Press, 2010). Specifically on books, see the numerous essays collected in Robert A. Gross and Mary Kelley, eds., *A History of the Book in America, Vol. 2: An Extensive Republic: Print, Culture, and Society in the New Nation, 1790–1840* (Chapel Hill: University of North Carolina Press, 2010); Scott E. Casper, ed., *A History of the Book in America, Vol. 3: The Industrial Book, 1840–1880* (Chapel Hill: University of North Carolina Press, 2007); Carl F. Kaestle and Janice A. Radway, eds., *A History of the Book in America, Vol. 4: Print in Motion: The Expansion of Publishing and Reading in the United States, 1880–1940* (Chapel Hill: University of North Carolina Press, 2009).

6 On the popularisation of science in the U.S., see Sally Gregory Kohlstedt, "Parlors, Primers, and Public Schooling: Education for Science in Nineteenth-Century America,"

Isis 81, no. 3 (1990): 424–45; Katherine Pandora, "Popular Science in National and Transnational Perspective: Suggestions from the American Context," *Isis* 100, no. 2 (2009): 346–58. Also see the introductory essay to the issue's focus section by Jonathan R. Topham.

7 Hans-Johann Glock, "The Anthropological Difference: What Can Philosophers Do to Identify the Differences Between Human and Non-Human Animals?," *Royal Institute of Philosophy Supplement* 70 (2012): 109. Also see Matthew Calarco, *Beyond the Anthropological Difference* (Cambridge: Cambridge University Press, 2020), esp. chapter 2; Peter Wiersbinski, "Conceiving the Anthropological Difference as a Categorical Divide," in *Empirical Research and Normative Theory: Transdisciplinary Perspectives on Two Methodical Traditions Between Separation and Interdependence*, ed. Alexander Max Bauer and Malte Meyerhuber (Berlin: De Gruyter, 2020), 317–36.

8 See Jacques Derrida, *The Animal That Therefore I Am*, ed. Marie-Luise Mallet, trans. David Wills (New York: Fordham University Press, 2008). In his last seminar, Derrida once again addressed the issue of this grotesque asymmetry of the distinction between human beings and the Animal. While the former "are supposed to belong to the same species or the same genus, the human species, the human race," the latter "designates a set with no other unity . . . than a negative one" – "not being a human being" – because there is "no other positively predicable unity between the ant, the snake, the cat, the dog, the horse, the chimpanzee – or the sperm whale." Jacques Derrida, *The Beast and the Sovereign, Vol. 2*, ed. Michel Lisse, Marie-Louise Mallet, and Ginette Michaud, trans. Geoffrey Bennington (Chicago: University of Chicago Press, 2011), 8.

9 Lewis Henry Morgan, "The Human Race," *The Nation* 15, no. 387 (November 28, 1872): 354.

10 Carrie Rohman, *Stalking the Subject: Modernism and the Animal* (New York: Columbia University Press, 2009), 5. On the influence and reception of Darwin in the U.S., see the essays in Jeannette Eileen Jones and Patrick B. Sharp, eds., *Darwin in Atlantic Cultures: Evolutionary Visions of Race, Gender, and Sexuality* (New York: Routledge, 2010); Tina Gianquitto and Lydia Fisher, eds., *America's Darwin: Darwinian Theory and U.S. Literary Culture* (Athens: The University of Georgia Press, 2014); also see George E. Webb, *The Evolution Controversy in America* (Lexington: University Press of Kentucky, 1994); Philip J. Pauly, *Biologists and the Promise of American Life: From Meriwether Lewis to Alfred Kinsey* (Princeton: Princeton University Press, 2000).

11 Carl N. Degler, *In Search of Human Nature: The Decline and Revival of Darwinism in American Social Thought* (New York: Oxford University Press, 1991), 6. For a comprehensive history of the concept of evolution and its reception, see Peter J. Bowler, *Evolution: The History of an Idea*, 3rd ed. (Berkeley: University of California Press, 2003).

12 Raymond Corbey, *The Metaphysics of Apes: Negotiating the Animal-Human Boundary* (Cambridge: Cambridge University Press, 2005), 34. On the Great Chain of Being, see Arthur O. Lovejoy, *The Great Chain of Being: A Study of the History of an Idea* (Cambridge: Harvard University Press, 1936); Peter J. Bowler, *Progress Unchained: Ideas of Evolution, Human History and the Future* (Cambridge: Cambridge University Press, 2021), chapter 2.

13 Raymond Corbey, *The Metaphysics of Apes: Negotiating the Animal-Human Boundary* (Cambridge: Cambridge University Press, 2005), 34, original emphasis.

14 Phillip Sloan, "The Gaze of Natural History," in *Inventing Human Science: Eighteenth-Century Domains*, ed. Christopher Fox, Roy Porter, and Robert Wokler (Berkeley: University of California Press, 1995), 141.

15 Phillip Sloan, "The Gaze of Natural History," in *Inventing Human Science: Eighteenth-Century Domains*, ed. Christopher Fox, Roy Porter, and Robert Wokler (Berkeley: University of California Press, 1995), 113, 114.

16 Phillip Sloan, "The Gaze of Natural History," in *Inventing Human Science: Eighteenth-Century Domains*, ed. Christopher Fox, Roy Porter, and Robert Wokler (Berkeley: University of California Press, 1995), 122.
17 Frank Luther Mott, *A History of American Magazines*, vol. 1: 1741–1850 (Cambridge: Harvard University Press, 1930), 341.
18 On the concept of human nature, see David L. Hull, "On Human Nature," *PSA: Proceedings of the Biennial Meeting of the Philosophy of Science Association* 2 (1986): 3–13; Grant Ramsey, "Human Nature in a Post-Essentialist World," *Philosophy of Science* 80, no. 5 (2013): 983–93; David Livingstone Smith, "Indexically Yours: Why Being Human Is More Like Being Here Than Like Being Water," in *The Politics of Species: Reshaping Our Relationships with Other Animals*, ed. Raymond Corbey and Annette Lanjouw (Cambridge: Cambridge University Press, 2013), 40–52; Maria E. Kronfeldner, *What's Left of Human Nature? A Post-Essentialist, Pluralist, and Interactive Account of a Contested Concept* (Cambridge: MIT Press, 2018); Matthew Calarco, *The Boundaries of Human Nature: The Philosophical Animal from Plato to Haraway* (New York: Columbia University Press, 2022).
19 Francis E. Brewster, *The Philosophy of Human Nature* (Philadelphia: Getz and Buck, 1851), v.
20 Martyn Paine, *A Discourse on the Soul and Instinct, Physiologically Distinguished from Materialism* (New York: E. H. Fletcher, 1849), 8, 9.
21 Henry Calderwood, "The Problem Concerning the Human Will, as Related to Scientific and Philosophic Theories," *The Princeton Review* 54 (September 1878): 329, 331, 333.
22 "Hints on Human Nature, Style, Etc.," *The Knickerbocker* 26, no. 6 (December 1845): 536.
23 Matthew Calarco, *Zoographies: The Question of the Animal from Heidegger to Derrida* (New York: Columbia University Press, 2008), 52.
24 Tim Ingold, "Humanity and Animality," in *Companion Encyclopedia of Anthropology*, ed. Tim Ingold (London: Routledge, 1994), 19.
25 John Quincy Adams, "Society and Civilization," *The American Review* 2, no. 1 (July 1845): 80, original emphasis.
26 John Quincy Adams, "Society and Civilization," *The American Review* 2, no. 1 (July 1845): 80, original emphases.
27 John Quincy Adams, "Society and Civilization," *The American Review* 2, no. 1 (July 1845): 80.
28 John Quincy Adams, "Society and Civilization," *The American Review* 2, no. 1 (July 1845): 81.
29 Louis Agassiz and Augustus A. Gould, *Principles of Zoology: Touching the Structure, Development, Distribution and Natural Arrangement of the Races of Animals, Living and Extinct* (Boston: Gould, Kendall and Lincoln, 1848), 43.
30 See Kay Anderson and Colin Perrin, "'Removed from Nature': The Modern Idea of Human Exceptionality," *Environmental Humanities* 10, no. 2 (2018): 447–72.
31 "Passions and Intelligence of Men," *The American Review* 16, no. 95 (November 1852): 541–51, here 541.
32 Myron Benton, "Instinct Demoralized," *Putnam's Monthly Magazine* 11, no. 5 (May 1868): 524; also see "Man and Monkey," *Scientific American* 10, no. 10 (November 18, 1854): 80; "Man and the Monkey," *Harper's New Monthly Magazine* 40, no. 235 (1869): 148–50. According to a writer in the *Scientific American*, the monkey was no less than "a standing libel upon the animal-man." "The Wonders of the Animal Kingdom," *Scientific American* 8, no. 10 (March 7, 1863): 148. In its more modern vein, the relevance of "simianness" for ideas about anthropological difference can be traced back to works like Edward Tyson's *Orang-Outang, sive Homo Sylvestris: Or the Anatomy of a Pygmie Compared With That of a Monkey, an Ape, and a Man* (1699). For a comprehensive historical discussion, see Corbey, *The Metaphysics of Apes*.

33 Augustus R. Grote, "Man and His Structural Affinities," *Popular Science Monthly* 13 (August 1878): 430–31.
34 Augustus R. Grote, "Man and His Structural Affinities," *Popular Science Monthly* 13 (August 1878): 434.
35 William Byrd, *The Westover Manuscripts: Containing the History of the Dividing Line Betwixt Virginia and North Carolina; A Journey to the Land of Eden, AD 1733; and A Progress to the Mines* (Petersburg: Edmund and Julian C. Ruffin, 1841), 83.
36 "Instinct and Reason," *Boston Investigator*, January 13, 1847, n.p.
37 "Mind in Animals," *The Ladies' Repository* 17, no. 3 (March 1857): 135, original emphasis.
38 Hans-Johann Glock defines the two poles of the debate as "lingualism" and "mentalism," the first denying that animals have thoughts in the absence of linguistic ability, the second arguing that "thoughts of animals differ from those of humans only in degree" and "are held to be in a language of thought, not in a public language." Hans-Johann Glock, "Animals, Thoughts and Concepts," *Synthese* 123, no. 1 (2000): 35–64, here 35. For a historical discussion, see John E. Joseph, *Language, Mind and Body: A Conceptual History* (Cambridge: Cambridge University Press, 2018).
39 William Cardell, *Essay on Language* (New York: C. Wiley, 1825), 9. For a counter perspective, see Edward Payson Evans, "Speech as a Barrier Between Man and Beast," *The Atlantic Monthly* 68, no. 407 (September 1891): 299–312.
40 Susan Pearson, "Speaking Bodies, Speaking Minds: Animals, Language, History," *History and Theory* 52, no. 4 (2013): 91–108, here 93.
41 "Our Speechless Slaves. Lecture by Henry Bergh," *Cincinnati Daily Gazette*, December 12, 1873, 2.
42 "Unknown Tongues," *Putnam's Monthly Magazine* 6, no. 32 (August 1855): 130–38, here 135.
43 "Unknown Tongues," *Putnam's Monthly Magazine* 6, no. 32 (August 1855): 130–38, 136, 135.
44 Bayard Taylor, "Studies of Animal Nature," *The Atlantic Monthly* 39, no. 232 (February 1877): 135–42, here 138.

Editorial Headnote

1. William Bartram, "Anecdotes of an American Crow," *Philadelphia Medical and Physical Journal* 1 (1804): 89–95.

William Bartram (1739–1823) was the son of John Bartram, a self-taught Pennsylvania naturalist of high international regard who served as royal botanist to King George III in the decade leading up to the Revolutionary War. William was a talented artist, and after a number of unsuccessful business ventures his fortunes changed when his drawings of plants were brought to the attention of English physician and plant collector John Fothergill by English merchant Peter Collinson, who shared Fothergill's interest in botany and had already financially supported the work of the older Bartram. Fothergill became a patron of William, awarding him a stipend that allowed him to fully devote himself to his work as a naturalist and realise the extensive trip through the American Southeast that would become the subject of his famous *Travels* (the book's full title is much longer), published in 1791. Primarily undertaken for the purpose of collecting plant specimens, Bartram's account also includes observations on the animal life and Native American communities of the region.[1]

In the following article for the *Philadelphia Medical and Physical Journal*, a periodical founded in 1804 by Pennsylvania physician and botanist (and friend of Bartram's) Benjamin Smith Barton (see B. S. Barton, Vol. 5),[2] Bartram shares his humourous observations of an American crow, a synanthropic species widely distributed across North America. However, Bartram is careful to preface his account with the caveat that "we do not here speak of the crow, collectively," only of Tom, an individual of the species, members of which Bartram speculates might "differ as widely as men do from each other." As indicated in the article's title, Bartram's interpretations of Tom's behaviour and "incontestible demonstrations of a regular combination of ideas, premeditation, reflection, and contrivance" rely on what psychologist Margaret Floy Washburn (see this section) would later criticise as the "anecdotal method" in studying animal minds. Still, Bartram's conclusions about Tom's "intellectual powers" seem to have been grounded in a long acquaintance with an animal he had personally "reared from the nest" and are lent credence by more recent studies on corvid intelligence.[3] (His meticulous description of Tom's antics in teasing a "large dog" by pulling the animal's tail, too, has been confirmed through broad social media video documentation of similar cases.)

Notes

1 Robert McCracken Peck, "Bartram, William," in *American National Biography*, ed. John A. Garraty and Mark C. Carnes, vol. 2 (New York: Oxford University Press,

1999), 297–99; Christoph Irmscher, *The Poetics of Natural History: From John Bartram to William James*, 2nd ed. (New Brunswick: Rutgers University Press, 2019), chapter 1.

2 See Myrl Ebert, "The Rise and Development of the American Medical Periodical 1797–1850," *Bulletin of the Medical Library Association* 40, no. 3 (1952): 243–76.

3 See, for example, Nathan Emery, *Bird Brain: An Exploration of Avian Intelligence* (Princeton: Princeton University Press, 2016). Carel ten Cate and Sue D. Healy, eds., *Avian Cognition* (Cambridge: Cambridge University Press, 2017).

1

ANECDOTES OF AN AMERICAN CROW (1804)

William Bartram

Source: William Bartram, "Anecdotes of an American Crow," *Philadelphia Medical and Physical Journal* 1 (1804): 89–95.

IT is a difficult task to give a history of our Crow. And I hesitate not to aver, that it would require the pen of a very able biographer to do justice to his talents.

Before I enter on this subject minutely, it may be necessary to remark, that we do not here speak of the crow, collectively, as giving an account of the whole race (since I am convinced, that these birds differ as widely as men do from each other, in point of talents and acquirements), but of a particular bird of that species, which I reared from the nest.

He was, for a long time, comparatively a helpless, dependent creature, having a very small degree of activity or vivacity, every sense seeming to be asleep, or in embryo, until he had nearly attained his finished dimensions, and figure, and the use of all his members. Then, we were surprised, and daily amused with the progressive development of his senses, expanding and maturating as the wings of the youthful phalæna,[1] when disengaged from its nympha-shell.

These senses, however, seemed, as in man, to be only the organs or instruments of his intellectual powers, and of their effects, as directed towards the accomplishment of various designs, and the gratification of the passions.

This was a bird of a happy temper, and good disposition. He was tractable and benevolent, docile and humble, whilst his genius demonstrated extraordinary acuteness, and lively sensations. All these good qualities were greatly in his favour, for they procured him friends and patrons, even among *men,* whose society and regard contributed to illustrate the powers of his understanding. But what appeared most extraordinary, he seemed to have the wit to select and treasure up in his mind, and the sagacity to practise, that kind of knowledge which procured him the most advantage and profit.

He had great talents, and a strong propensity to imitation. When I was engaged in weeding in the garden, he would often fly to me, and, after very attentively observing me in pulling up the small weeds and grass, he would fall to work, and with his strong beak, pluck up the grass; and the more so, when I complimented

DOI: 10.4324/9781003032717-2

him with encouraging expressions. He enjoyed great pleasure and amusement in seeing me write, and would attempt to take the pen out of my hand, and my spectacles from my nose. The latter article he was so pleased with, that I found it necessary to put them out of his reach, when I had done using them. But, one time, in particular, having left them a moment, the crow being then out of my sight, recollecting the bird's mischievous tricks, I returned quickly, and found him upon the table, rifling my inkstand, books, and paper. When he saw me coming, he took up my spectacles, and flew off with them. I found it vain to pretend to overtake him; but standing to observe his operations with my spectacles, I saw him settle down at the root of an apple-tree, where, after amusing himself, for awhile, I observed, that he was hiding them in the grass, and covering them with chips and sticks, often looking round about, to see whether I was watching him. When he thought he had sufficiently secreted them, he turned about, advancing towards me, at my call. When he had come near me, I ran towards the tree, to regain my property. But he, judging of my intentions, by my actions, flew, and arriving there before me, picked them up again, and flew off with them, into another apple-tree. I now almost despaired of ever getting them again. However, I returned back to a house, a little distance off, and there secreting myself, I had a full view of him, and waited to see the event. After some time had elapsed, during which I heard a great noise and talk from him, of which I understood not a word, he left the tree, with my spectacles dangling in his mouth, and alighted, with them, on the ground. After some time, and a great deal of caution and contrivance in choosing and rejecting different places, he hid them again, as he thought very effectually, in the grass, carrying and placing over them chips, dry leaves, &c., and often pushing them down with his bill. After he had finished this work, he flew up into a tree, hard by, and there continued a long time, talking to himself, and making much noise; bragging, as I supposed, of his achievements. At last, he returned to the house, where not finding me, he betook himself to other amusements. Having noted the place, where he had hid my spectacles, I hastened thither, and after some time recovered them.

This bird had an excellent memory. He soon learned the name which we had given him, which was Tom; and would commonly come when he was called, unless engaged in some favourite amusement, or soon after correction: for when he had run to great lengths in mischief, I was under the necessity of whipping him; which I did with a little switch. He would, in general, bear correction with wonderful patience and humility, supplicating with piteous and penitent cries and actions. But sometimes, when chastisement became intolerable, he would suddenly start off, and take refuge in the next tree. Here he would console himself with chattering, and adjusting his fathers, if he was not lucky enough to carry off with him some of my property, such as a pen-knife, or a piece of paper; in this case, he would boast and brag very loudly. At other times, he would soon return, and with every token of penitence and submission, approach me for forgiveness and reconciliation. On these occasions, he would sometimes return, and settle on the ground, near my feet, and diffidently advance, with soft-soothing expressions, and a sort of circumlocution; and sit silently by me, for a considerable time. At other times, he would confidently come and settle upon my shoulder, and there solicit my favour and pardon, with soothing expressions, and caressing gesticulations; not omitting to tickle me about the neck, ears, &c.

Tom appeared to be influenced by a lively sense of domination (an attribute prevalent in the animal creation): but, nevertheless, his ambition, in this respect, seemed to be moderated by a degree of reason, or reflection. He was, certainly, by no means tyrannical, or cruel. It must be confessed, however, that he aimed to be master of every animal around him, in order to secure his independence and his self-preservation, and for the acquisition and defence of his natural rights. Yet, in general, he was peaceable and social with all the animals about him.

He was the most troublesome and teasing to a large dog, whom he could never conquer. This old dog, from natural fidelity, and a particular attachment, commonly lay down near me, when I was at rest, reading or writing under the shade of a pear-tree, in the garden, near the house. Tom (I believe from a passion of jealousy) would approach me, with his usual caresses, and flattery, and after securing my notice and regard, he would address the dog in some degree of complaisance, and by words and actions; and, if he could obtain access to him, would tickle him with his bill, jump upon him, and compose himself, for a little while. It was evident, however, that this seeming sociability was mere artifice to gain an opportunity to practise some mischievous trick; for no sooner did he observe the old dog to be dozing, than he would be sure to pinch his lips, and pluck his beard. At length, however, these bold and hazardous achievements had nearly cost him his life: for, one time, the dog being highly provoked, he made so sudden and fierce a snap, that the crow narrowly escaped with his head. After this, Tom was wary, and used every caution and deliberation in his approaches, examining the dog's eyes and movements, to be sure that he was really asleep, and at last would not venture nearer than his tail, and then by slow, silent, and wary steps, in a sideways, or oblique manner, spreading his legs, and reaching forward. In this position, he would pluck the long hairs of the dog's tail. But he would always take care to place his feet in such a manner as to be ready to start off, when the dog was roused and snapped at him.

It would be endless (observes my ingenious friend, in the conclusion of his entertaining account of the crow) to recount instances of this bird's understanding, cunning, and operations, which, certainly, exhibited incontestible demonstrations of a regular combination of ideas, premeditation, reflection, and contrivance, which influenced his operations.

Editorial Endnote

1 *phalaena*: a genus of moths in the original taxonomic system developed by Swedish naturalist Carl Linnaeus (1707–1778) in his *Systema Naturae* (1758).

Editorial Headnote

2. Frederick Augustus Rauch, *Psychology; or, a View of the Human Soul; Including Anthropology.* New York: M. W. Dodd, 1841.

Prior to his emigration to the United States in 1831, Frederick Augustus Rauch (1806–1841) had studied classical literature at the University of Giessen, Germany, with the intention of pursuing a career in philology. But after a series of professional and legal disputes involving his doctorate and subsequent habilitation had likely poisoned all prospects of a career in German academia and also left him in serious debt, Rauch decided to leave his home country for Pennsylvania, where he managed to secure a position at the German Reformed Seminary in York, serving as principal of the classical department and Professor of Biblical Literature. In 1836, the school, now located in Mercersburg, became Marshall College, with Rauch as its first president and Professor of Hebrew, Greek, and German.[1]

When Rauch died a premature death at age 34 in early March 1841, he was in the process of preparing for publication the second edition of *Psychology*, the major work of his short career. Informed by the philosophy of Georg Wilhelm Friedrich Hegel, Rauch's was one of the first American books to be published with "psychology" in its title, though at this time the latter was still very much a subject of philosophical inquiry and yet to develop into a discipline of its own. With its focus on the study of the human "soul," it was also closely tied to theology – the "science" of religion, according to Rauch, which shared with psychology a "common subject, which is *man*" (2). Without religion, "man" was "only an animal, more cunning, crafty and prudent, than all the others, one that can invent machines, but he is no longer the lord of the earth, the image of his Creator" (2).

Rauch's book is divided into an introduction and two main parts that cover, respectively, "anthropology" (dealing primarily with the nature of the mind and mind-body relations) and "psychology" (dealing primarily with "reason" and "will" and concluding with a section on religion). The following excerpts are from the first two chapters of the book, which make up its introduction and establish the argument for human uniqueness, insisting that the human being was both physically and mentally "no more a mere continuation of the animal, than the animal is merely a continuation of the vegetable."

Note

1 James A. Good, "Rauch, Frederick Augustus (1806–1841)," in *The Dictionary of Early American Philosophers*, ed. John R. Shook, vol. 2 (New York: Continuum, 2012), 861–65.

2

PHYSICAL AND PSYCHOLOGICAL DIFFERENCES BETWEEN HUMANS AND ANIMALS (1841)

Frederick A. Rauch

Source: Frederick Augustus Rauch, *Psychology; or, a View of the Human Soul; Including Anthropology* (New York: M. W. Dodd, 1841), 9–17, 30–44.

DIFFERENCE BETWEEN MAN AND THE ANIMAL.

It is a very general remark, that man is the highest order of animals, or that he is an animal gifted with reason. Were this correct, we might say with equal truth, that the animal is man without reason; or that some of the plants, which seem to form a transition from the vegetable to the animal kingdom, are animals without sensation. Though man has, physically speaking, many things in common with the animal, he nevertheless differs wholly from it. Man is no more a mere continuation of the animal, than the animal is merely a continuation of the vegetable. The difference is perceptible, both *physically* and *psychologically*.

1. Man has the same *physical* functions that the animal has, but they are more perfect and more delicate. The body of the animal is either covered with scales or feathers, with fur, wool, or bristles. All of these approach more or less nearly to the nature of vegetables; and some of them, as for instance, the shell of snails and the tortoise, to inorganic nature. These insensible substances, interposed between the skin and the elements that surround and affect living beings, deprive animals of the more tender and delicate sensations, at the same time that they like conductors protect them from the inclemency of the seasons. The body of man is covered with a thin, highly sensitive, and beautiful skin, which is not concealed under a vegetable and lifeless veil, but unveiled exhibits the presence of the blood, the pulsations of the heart, and the utterance of animation over the whole body. Everywhere sensitive, the swelling life shines forth from within, and the fresh, blooming color of the skin is but the expression of the internal life, the mirror of its healthy or sickly state. Over the face of man are shed beauty and spirit, and even animals are said to gaze at him with mute wonder. The complaint, that man alone is born without raiment, is silly, for this apparent helplessness is the

DOI: 10.4324/9781003032717-3

source not only of the most various and delicate sensations, and knowledge derived from them, but also of much ingenuity. Man has not claws, but hands, that are susceptible of many different positions, by which he handles the chisel, which pours life and beauty over the hard, cold marble, the pencil that animates the canvas, the instrument from which he draws forth sweet melodies, and the iron, from which he forms the weapons denied him by nature. His body is so fashioned that he must walk upright; for while no animal is intended to walk otherwise than on all-fours, the proportions of the human frame are such, as to render any other than an upright position almost impossible. His legs are much longer than his arms, his knees bend forward, his eyes are in front and not at the sides, the ligaments of the neck are weak and incapable of supporting the head when hanging down, the arms are at a great distance from each other, and the breast is broad and full. A horizontal position would drive the blood into the head with such violence as to cause stupor. Man is made to turn his head from the earth to the sky, from the right to the left, to view now the crawling insect beneath his feet, and now the millions of stars above his head. To the fish it is natural to swim, to the bird to fly, to man to walk upright. The Greek word for man (ἀνθρωπος) indicates the difference between man and animals in this respect. It influences our whole being and nature. Even the bees, when they have lost their queen-bee, cause the larva of a future laboring bee to be transformed into a queen by changing its horizontal to an upright position, and giving it different food.

The same superiority is visible in the human face, its proportions and features. With the animal, the most prominent part is the mouth; with man, the upper part of the face. With the animal, the mouth, jaws, and teeth are to serve only physical purposes; they are formed to pluck the grasses and twigs, or to seize and carry their prey, and thus to perform at the same time the service of the hands of man. The human mouth, with its beautiful and sensitive lips, its regular rows of teeth, serves not only the body, but the soul; nourishes not only the stomach, but also the understanding. Its muscles are so movable, that according to Haller's calculation,[1] it may pronounce in one minute fifteen hundred letters. The contraction of a muscle forming the letter must consequently take place in the three thousandth, and the vibrations of the stylo-pharyngean muscle[2] in pronouncing a letter, in the thirty thousandth part of a minute. "No bird flies as fast as the winged words fall from the lips of man."[3]

Comparing the animal physiologically with man, we cannot but perceive a great difference in this respect also. The lower the animal in the scale of being, the more it is confined to one and the same food. To this food it is directed by instinct, by constitution and appetite, as the magnet to the pole. Man, on the contrary, selects his food and drink from all the kingdoms of nature, from the salt of the ocean to the mushrooms of the forests, from the oyster and the amphibious turtle to the lofty sailers of the air and ibex of the High Alps. He prepares his food by fire, and the story of Prometheus is not a mere fable. The laws of his digestion differ widely from those of animals, the functions of assimilation penetrating more thoroughly the elements of nutrition. The flesh of the fish, when compared with that of the bird, is found less formed and solid; that of the bird less so than that of the quadruped, but the flesh of man is more perfect than that of any animal. This shows itself externally on the whole surface of the human body, and especially in the "morbidezza"[4] of the skin, the trial-point of all artists.

2. The *psychological* difference between man and animals is yet more striking. The animal has in common with man, Sensation and Perception. By sensation we understand an internal motion or activity, produced in a sensitive organ by something external. The organ may be seen externally, but this internal activity or motion cannot be observed; it can only be felt. The eye, for instance, is the organ of sight; the fluid surrounding it is constantly in motion, and this motion may be seen, because it is an external one. But when the light falls upon the eye, it causes a sensation, which as an internal motion, is invisible to the eye of the observer. He may notice the dilation or contraction of the pupil, but cannot see sight itself, nor hear hearing. Comparing the vibrations of a string of the piano with those of a nerve, subject to sensation, we shall find the former altogether mechanical and external, the string moving away from itself. The nerve, when affected, trembles within itself, and self-touched in its motions, it has sensation, or feels. The vibrating string of the piano gives a sound, but the sound is not felt by it; the nerves of the ear, on the other hand, receive a sensation from it. These trembling motions of the nerves are called *sensations*, because they are peculiar to the senses. Now, the animal has these sensations in common with man, but with this difference. In the animal *one* sense prevails over all the others, and these are subservient to it. In the eagle, for example, it is the eye that is predominant; from immense heights he observes the mouse creeping on the soil, and darts upon it, certain of his prey. Yet this one sense has always reference to the means of subsistence, which the animal seeks under its guidance; so that while it may be extremely acute and successful in discovering the food, it may be dull and stupid in respect to other objects, of which man receives the most accurate sensations through the same sense. The ear of the wood-cock is acute in perceiving any rustling noise, as that in the fallen leaves of the forests, but shrill and clear sounds it does not notice. The eye perceives only such objects as reflect the light upon it. The eye of man may, however, direct itself to the different parts of these objects, to their color, proportions, size and form, motion or rest, and inspect each by itself, while the eye of the animal, for want of reason, is forced to admit a sensation from all these parts at once, and consequently receives but a confused impression. As one sense prevails in animals, the others are found less active. The lion has an excellent scent, but his sight is weak. Hence the animal is under the dominion of one sense, while the harmonious and equal strength of all the senses places man above them, and makes him master of

them all. Which of the senses prevails in an animal, depends always upon the species to which it belongs. In the eagle it is sight, in the mole hearing, in the vulture scent; but wherever one sense predominates, the others must subserve and be directed by it. In man, no sense being more acute than another, none reigns, but all are co-ordinate with each other, and subordinate to the understanding. The animal is *vis sentiens*, man *natura intelligens.*[5] Hence it is, too, that we do not speak of *insanity* or *derangement*, but of *madness*, when animals can no longer distinguish one object from another, but like infuriated elephants or horses, trample under their feet every object that opposes their course. And, finally, man sees not only with his eye external objects, but is also able to examine the eye, by which he sees; the animal can neither see its eye nor itself.

Perception, however, is more than Sensation. The latter is and remains in contact with the object, by which it is called forth, and is consequently dependent on it. The sensation of hearing is impossible without the vibration of the air; that of seeing is impossible without the presence of light. Sight and light, hearing and sound, are so inseparable, that the one could not be without the other. Without light the eye could not see, without an eye, there would be an eternal night. And if there were no ear, the brooks might continue to murmur, the waving trees to rustle their branches, the winds to roar, but to earthly beings nature would be silent as the grave. Nor can man avoid admitting a sensation, when the element that excites it, acts upon the organ. As we cannot have a sensation of light unless it falls upon the eye, so we must have it, when the eye is affected by it. We cannot taste salt unless it lies on the tongue, but when once it is brought in contact with the tongue, we must have such a sensation, as its specific nature is capable of exciting. In sensation, therefore, we depend wholly on the presence of external objects, and are determined by them. But after we have once had a sensation, we may have perceptions of the objects of sensation, and these perceptions are possible without the presence of the objects. This appears from the fact that one, who in the latter part of his life becomes blind, may have perceptions of all he once saw. Yet perceptions are impossible without previous sensations, for one born blind can have no perception of color or form. Another distinction between sensation and perception is this. Sensation always exists in an organ; perception not. When I hear a fine melody for the first time, I have a sensation of it; but when afterwards, without hearing it, it floats in my mind, I have a perception of it. Or when I experience hunger, I may have a perception of food, though it be not present, but when I eat, I have a sensation of it. The animal has perceptions as well as man. The hunter's dog dreams, and pursues in his dream the hare or the stag. The dog, when near to his master, has a sensation of him by scent, or sight; but when seeking him for days in succession, he can have only a perception of him. The animal is confined within the sphere of sensation and perception, and as its sensations are limited to its natural wants, so must be its perceptions. The perceptions of man are as much more numerous and accurate, as his sensations are more various and acute. But he has in addition to sensations and perceptions, what the animal has not, which we may express by the term *apperceptions*.[6]

As the animal is separated by sensation from the plant, so man is elevated by apperception above the animal. Perception and apperception differ widely. The objects of perception are always such as are single and met with in a certain place and time. The eagle, that builds his nest on a high rock, has not the least idea of the nature of the stone on which his nest is placed, nor of the region in which it stands, but he carries with him the image of this single rock, as it stands in a particular place, and noticing no resemblance between it and other rocks, he would find it among thousands of others, its peculiar features being strongly and solely impressed upon his eye. The objects of apperception, on the other hand, are the kind, the species and individuals of things. It is by apperception, that man distinguishes between his perceptions and the objects perceived; and again, that he classifies nature and its productions, by discovering the union in the greatest variety, as when he says: these bushes are rose-bushes; and by distinguishing one class from another, as pears from apples, and these from peaches. This the animal can never do. It sees, it perceives the grass, but it never arranges it according to its botanical classes. The dog universally will pursue the hare, and it may seem that he does so, because he knows this class of animals, and distinguishes between it and others; but the truth is, that all hares being exactly alike in size, form, and scent, will produce the same sensations in the dog, and these will always set him in motion. This then is the broad difference, the chasm between man and the animal: the former can *think*, the latter cannot, for it is *glebae adscriptus*.[7] It lives, but acquires no experience; it eats its food daily, but never knows what this food is. Some indeed have gone so far as to say, that animals not only judge, but draw conclusions from causes to effects. To draw conclusions is the highest power of human reason, and if they could do this, they would be able to *think*, and to *will* like man, and to have apperceptions like him. Such men cannot know what thinking is, or what is to be understood by judging and drawing conclusions. Animals have no idea of power, of capacities, of energy, of proportions, of beauty, of truth, and consequently none of cause and effect. None of these are visible or accessible to the senses, but only to thought, which it yet remains to be proved, that animals possess. The process by which we arrive at conclusions is simply this. We have three thoughts; each differs from the other, and each is included within a certain limit. But while all differ, one is capable of uniting the two others, and of removing this difference. Thus two thoughts are reduced to one class by a third one connecting them. To make use of an example frequently adduced, the dog once whipped, fears as soon as he sees his master take the cane with which he has been beaten on a former occasion. The cane is one thing, the master's intention another, and the pain proceeding from the whipping, a third. Now it would be foolish in the extreme, to say, that the dog connects the idea of pain with the cane by the intention of the master. He has but a *confused* impression, and without any conclusions or judgments, he darkly connects things as they formerly were connected, and anticipates consequences, without being conscious of such a connection, or without having any thing like an idea of cause and consequence. Hunters, it is true, have many anecdotes about the acuteness and ingenuity of animals, as do sailors voyaging to distant countries. The fact, however, is, that unless their game were bound by the invariable laws of instinct in all its actions, unless one fox would dig its hole as the other, and all stags would go to the water, and seek food at regular hours, and live in certain places at different seasons, the hunters would not be able to entrap them.

Animals, finally, cannot have any emotions; neither joy nor grief, neither hope nor fear. The external expressions of these emotions are *weeping* and *laughing*, neither of which has as yet been observed in animals. We all remember from Homer's Iliad, that when the noble Patroclus, alone and at a distance from his true friend Achilles, fell by the hands of the Trojans, his horses shed big tears and refused to obey, because they missed the well known voice of their beloved master. But these tears belong to poetry; they are the tears which Homer himself wept at the death of the hero his fancy had created. So poetry attributes innocence to the lily, because it is of the purest white; modesty to the violet, because it blooms and exhales its fragrance unseen; love to the rose, because the cheeks of the maiden blush like it, when for the first time she feels this noble emotion. What is joy in man, springing from a feeling that connects itself with some thought, in animals, is but a physical sensation or bodily pleasure, the agreeable re-action of the muscles against some external influence, the satisfaction of some want. And so what is grief in man, is but a physical pain, or suffering in the animal. The dog, that lays himself upon the grave of his master and remains there until he dies, does so not from deliberation and free choice, but being forced by the chain of habit, which he is unable to break. Nor do animals fear or hope, for neither the future nor the past is known to them. A dark anxiety which they do not understand, a confused anticipation, is all of which they are susceptible.

INSTINCT.

What comparative anatomy is for the study of the anatomy of man, that, instinct and an investigation of its nature is for the study of Mental Philosophy. Within its sphere we discover phenomena that are full of design and calculation, and yet reason being wanting, will and self-consciousness being entirely absent, we cannot attribute these designs to animals, but must ascribe them to him who works by eternal laws through their instinct. The physiologist finds it necessary, in order to understand fully the different organs of man, to compare them with those of animals. He must trace them from their first indications in the lower classes of animals, through all their different gradations up to man. He thus will discover the importance of each part, by perceiving what degree of sight or hearing those parts of the eye or the ear afford, which are met with in animals that have not those organs perfectly formed. So it is useful to the Mental philosopher to cast a glance upon the development of psychological powers in nature below man. There are animals that consist of a single organ, which, while in its connection with a complicated organism in man, it is subordinate to higher ones, is the only one in them or at least prevails over the few others that they may possess. In viewing such a being, we may see what kind of life a single organ is capable of producing, and what its share must be in the constitution of man. So we know, that in muscles and snails, the liver and heart alone are fully formed; in many insects the wind-pipe; in others the lungs; in the polypous the stomach; in infusoria the gut. Nature, according to this view, contains all the parts of man, but not as man has them. In the lowest animals, single parts are sufficient to form the whole being; more of them become united at the higher stages of animal life, until finally all appear well-proportioned in man, the tree, whose leaves are scattered throughout nature; and as a machine can be known only when its parts are viewed singly, so man can be understood only, when we are acquainted with the inferior tribes in nature, which present to our inspection the different parts composing his system. Thus it is likewise with the psychological life of man. In viewing the nature of instinct we may see what kind of mental life sensation and perception, independent of reason and will, are able to produce, and thus we may learn how to value reason and will as we should. It is not the identity, but the difference, not the sameness but comparison, from which we may learn most, and whenever in a science we have gained a prominent point adapted to exhibit these differences, it will be well for us to pause for a moment over it. Such points are for science, what mountains are for travelers, who desire to observe a country. They will see more, when standing on mountain-tops, than any where else. The nature of plastic power in the vegetable world, and that of instinct in animals, will accordingly teach us more concerning Psychology, than any other portion of human science.

The different general phenomena of instinct have been arranged by the celebrated Reimarus,[8] in no less than fifty-eight different classes. It will be sufficient for the present purpose, to mention only a few. Animals, from the time of their birth, move with perfect skill from one place to another, and use all their limbs in a perfectly correct manner, and for the right purpose, without having received instruction. So the squirrel uses its forepaws at once properly; so the fish swims without teaching. Amphibious animals will move from one element into the other; and birds, insects, fishes, and even quadrupeds will seek, and unerringly find, distant countries, in order to enjoy the degree of heat or cold favorable to their constitutions. Other animals bury themselves when winter approaches. All animals select their food not only skillfully, but also seek it in the proper places, at the proper season, and at the proper time, by day or night; many are extremely cunning in catching their prey, and in laying up provision for inclement seasons; others know how to heal their wounds, how to erect dwellings for themselves, how to defend themselves from their enemies either by houses, as the beaver, or by regular wars, as some species of ants. All these phenomena of instinct, however, may be reduced to three great classes, one of which will comprehend all those that have reference to the *nutrition*, another those referring to the *motion*, and the third those relating to the *propagation* of animals, for the end of all instinct is the preservation of the *individual* and of the *race.*

Instinct pre-supposes, what is not found in the sphere of vegetation, sensation and perception; and while the plastic power of the vegetable kingdom extends also to the animal, instinct is confined to such beings as can feel. The plant grows and ripens, but it would be improper to say, except poetically, that it sleeps, or that it is fatigued, hungry and thirsty. It is true, that plants hang their leaves when they suffer from want of moisture; that, like the *lotus ornithopedioides*,[9] which folds its flowers at night and opens them again in the morning, they seem to sleep, yet when all is well investigated, we shall discover that we speak but metaphorically of the sleep of plants. For as they are never awake, so they cannot properly be said to sleep. But hunger and thirst, a tendency for motion and rest, and for the propagation of the race, are the peculiar phenomena of instinct, which we shall now investigate.

It is natural to every living being to sustain itself by food. But neither man nor animals would think of taking nourishment, did not the system and operation of digestion force them to do so. When by exercise or atmospheric influence, digestion is regularly promoted, the stomach will become empty, and the gastric juice will gather. The power of this juice will seek something to act upon, and finding no food, it will attack the coats of the stomach. If no food is administered, the stomach will make an attempt by contractions to remove this juice, and not succeeding in this, *death* is inevitable. It is the nature of instinct, – 1st, To feel the pain thus caused by the activity of this gastric juice, and to feel the danger of destruction. This feeling itself is of course painful, and is generally called *hunger.* – 2d, Instinct, as hunger, will impel the animal to attack the world around and seek for food. This appears already from the connection of hunger and an irresistible tendency to motion. The horse stamps when hungry, and were it not chained, it would go in search of food. The boa constrictor is constantly active when in want of food, but as soon as its hunger is satisfied, it lies sluggishly down and may be chased by a child. – 3d, Instinct will direct the animal to its proper food, and no sooner is this perceived by the particular sense that prevails in the animal and stands in the service of instinct, than a prophetic feeling of pleasure will at once induce the animal to seize upon it. Hence it is, that the sheep, without choice or consideration, will select salt from amongst arsenic, which would be impossible to man. Instinct then is, on the one hand, a feeling of want, and on the other, a feeling of the sympathy existing between this want, and the objects by which it is to be satisfied. This sympathy expresses itself in the animal by the pleasure it feels on perceiving its proper food.

[. . .] Animals are certainly not mere machines, as Descartes[10] maintained, but neither are they thinking beings, as many sensualists would like us to believe. Their life is confined to sensation and perception, and all their activity proceeds not from *will*, but from a feeling of pain or pleasure.

OF THE INGENUITY OF ANIMALS.

I hope, however, to throw still more light upon this interesting topic, by considering some productions of animals which seem to manifest ingenuity. Of this kind, are all those that on the one hand, answer as *means* for a certain *purpose;* for example, the web, by which the spider catches his prey, flies and insects, as skillfully as the fisher entraps in his net the inhabitants of the rivers. Wherever we perceive an adaptation of means to the end, there we allow ingenuity to be active. On the other hand, they must be separate from the animal that produced them. The shell of the snail, that of the tortoise, and of the armadilla, are very artist-like and beautiful, but they are formed by mere excretion, and by the influence of the atmosphere, and constituting parts of the animals themselves, they do not belong to the class of productions of which we speak. We do not think of ingenuity when we admire the beautifully colored wings of the butterfly, or when we delight in viewing the regular and beautiful formations of leaves, of buds and flowers, for all of them form parts of the beings in which we discover them, and are the products of the same plastic power which formed the animals or plants themselves; but when we see the cell of the bee, the larva of the caterpillar, or the nest of the tailor bird, we are at once struck with their ingenuity. When we examine the cover in which the chrysalis of a caterpillar awaits its future transformation, we find it full of design. Some of these coverings have a crown on one end, made of erect and stiff threads that form in the inside a smooth and comfortable surface, but offer stiff knots and points on the outside, so that they easily yield to a pressure from within, but make it difficult to be pressed in from without. Here is design, here is preparation for a change which the animal has to undergo but once in its life, and as we cannot feel willing to ascribe these phenomena to the knowledge and will of the animal, we must attribute them to a peculiar modification of instinct. This we feel ourselves the more strongly urged to do, when we ascertain that this tendency for artificial productions is not met within the more perfect animals which possess all the senses in considerable perfection. The elephant does not show a trace of it; the horse, the reindeer, the ibex, the zebra, are not possessed of it. But the tortoise, a stupid, sluggish, and awkward creature, paddles to the shore, digs with her clumsy feet a hole, and after having deposited her eggs in it, she covers it, levels it with the soil, and creeps several times over it, so that not a trace of the hole can be discovered. From this it will appear, that this modification of instinct is not found every where in the animal world, especially not in that part which by its completeness of the senses approaches most nearly the intellect of man, and the question offers itself, – Where is it met with?

It is not found among such animals as maintain a decided independence of the elements in which they live; such as have five senses, as the buffalo, the bison; nor again, among such as depend almost wholly on the element surrounding them. It is consequently confined to those, in which dependence on certain elements, equals their independence of them. Their independence is exhibited in an activity resulting from their wants and necessity; their dependence on the element is seen in the fact that they are wholly penetrated by it, and that in receiving it in so full a manner, they receive also its laws, and come under their influence. The animal has feeling, sensation, and perception, and while by its activity it produces something wholly different from itself, a production that cannot coalesce with itself; it is, at the same time, active in accordance with the laws of the element pervading its system, and hence the thing produced will receive its form, not less by the activity of the animal, than by the laws of its element. These works are therefore the joint products of the animal, and of those laws which in a lower sphere call forth the many forms of crystals, and in a higher one, the beautiful formations of plants, of buds and flowers, and which, not directly, but through the animal, form here the cell of the bee, and there the pyramid of the ant. As the idea expressed here is somewhat difficult to be understood, we may be permitted to give it in another form. When the plant grows and produces beautifully formed buds and flowers, we admit that it is the plastic power in the plant, which produces the flower. So likewise it is the plastic power of nature which produces the artificial works of insects. The animal has sensation; it has wants and feels them; this feeling urges it to be active, in order to remove the want. In so far, the animal. is not under the influence of the laws of *general*, but under those of its own specific nature. At the same time, it being under the influence of the laws of the element in which it lives, water, or air, or light, is determined in the manner of its activity, no less by these laws than by its own wants, so that its productions receive their *form* from the influences of nature upon them.

The most beautifully formed crystal does not fill us with as much wonder as the cell of the bee, merely because the former is considered by us as the direct product of nature, but the latter as that of the design and calculation of the animal. And yet both rest on the same general ground, with only this difference, that the one proceeds directly from the laws of nature, and the other through the medium of a living individual. [. . .]

It may be instructive to compare these artificial productions of animals with the works of human art. The great contrast between them will show the true nature of the former.

1. Animals are born not only with the capacity, but with the ready skill to produce artificial works. This is manifest from the fact, that these little creatures not only execute these works immediately after their birth, but also in the same way throughout their lives, without in the least improving them. The spider feels a tendency to weave his web, before he has seen the flies, to ensnare which he spins the thread. The ant-eater can scarcely yet move, when his nature already impels him to prepare the funnel, for the purpose of catching ants and other insects. To say that they learned this from seeing their parents do the same, would be contrary to experience. For, as Aristotle remarks, if we have three eggs hatched by artificial heat, one of a bird, one of a duck, and one of a serpent, we shall see the young bird try to fly, before its wings are grown; the duck to swim, and the serpent to creep into the earth, before they have seen any one of their kind do the same.[11] What man does, he must have learned by trial, but the caterpillar has only once in his life to undergo a transformation, and yet he knows how to spin a covering that will suit his future state, of which he cannot have the least idea. The work, to be produced, seems to bear a prophetic character, for while the larva is still of a cylindric form, he weaves a covering fitted to the form of the chrysalis, as if he had his future state before his eyes. Man is not born with any ready skill, like the animal; his arm allows the mere possibility of performing thousands of different operations, but

this possibility must be exercised and developed. Exercise demands both time and repetition, and produces experience; but experience is impossible without reason and judgment. All the artificial productions of animals are based on instinct; those of man on reason, will, and consciousness, and hence it is that the former have reference only to physical wants, but the latter to intellectual.

2. Some of the animals construct their artificial works from materials that are contained in themselves. The spider draws out the thread from an excrescence on its body, attaches it to some external object, and separates it from itself. The aquatic spider, that cannot live under water without air, draws forth from its nipple a moist substance, a kind of varnish, covers itself with it, and bursting this bladder, it forms by degrees a diver's bell of it, as large as half a pigeon's egg; by a few threads it fastens it to some solid object in the water, its opening hanging downward, and then filling it with air, it may sit in it below the water for a long time, and watch its prey. So the bee gains honey and wax by digestion, and forms its cells of them. But other animals, like man, use inorganic materials; the beaver takes wood, the bird twigs, though the latter lines his nest with hair, straws, etc. The house of the beaver, and that of man, may consist of nearly the same materials. There is, however, this difference. The animal uses these inorganic substances without knowing their qualities, without having the least idea of the powers that fit them for the uses to which they apply them. They do not use them, therefore, from choice or consideration, but being directed to them by instinct. The swallow builds its nest of the same materials now, which it used in the time of Pliny.[12] Man, on the other hand, knows the powers of different substances, and their fitness for various purposes; he, therefore, selects and judges. Stones may be good for one, wood for another building. He has works written on the different building materials, and architects must be well acquainted with the nature of timber and stone.

3. Animals produce all their works by their own limbs, for they have no instruments. The beaver saws the wood with his teeth, drives stakes into the ground, and smooths the earth with his tail. These limbs are so constructed that their natural motions have a tendency to produce the works, which appear to us so remarkable. The proboscis of the bee is well shaped to be thrust into the nectar-cup of buds, and to drink in their sweet juices. The German rat has a bag below the chin, into which it gathers grain; the mole has protruding fore-feet to dig with; the fish has fins and a broad, upright, standing tail, to swim. But man, according to Franklin,[13] is an animal that can invent machines. The savage, fastening a sharp stone to a club, uses it as an axe; he does not defend himself by his limbs, but by the bow, the arrow, and the tomahawk. Man has invented the saw and the hatchet, the sword and the gun, the furnace and the mill, and is daily increasing the number of machines. There is no animal that ever attempted to fabricate its works by the use of self-invented machines, or by fire. Monkeys, that are generally admitted to be ingenious, never think of keeping up the fire by adding wood, though they evidently delight in its warmth; nor of firing woods, or of destroying the property of those that injure them, or of preparing instruments of iron by the aid of fife.

4. The arts of animals have no history, as have those of man. They have no history in the *objective* sense of the word; they have no gradual development, no cultivation, no improvement; they are stationary, they are the same now that they always were. The arts of animals cannot have a history in the *subjective* sense of the term history, for there can be no historical narration where nothing is to be narrated and animals cannot be conscious of an improvement, when there is none. But it is the character of history, that on the one hand it develops gradually all the capacities of a nation, or of our race; and on the other, makes us conscious of them by narrating them; and hence it is, too, that history objectively means the actions related, *res gestae*, and subjectively the relation of them, *historia verum gestarum,* On the other hand, there was a time when man knew of no arts, not even of such as are now considered indispensable. But animals have always had the same arts in the same degree of perfection and for the same purposes as at present. Man had to invent his, and he changes, and improves them daily. The arts of men differ in different regions, while the same animals will always produce the same works, wherever they are. And so men of different capacities in the same region, will have different success, while animals of the same kind will have the same.

5. Finally, the purpose for which animals produce works of art, is limited to the sustenance of individual life and the preservation of the race. Hence their works are few in number. The spider can only spin a web; the bee can only build a cell. Hence it is too, that the bird does not build his nest before the time of propagation. The purposes, for which man exercises his ingenuity and inclination to art, are manifold. By mechanical art he invents instruments, to serve his desire for knowledge. The press, the paper, the ink, the watch, the electrical machine, the compass, the telescope, give sufficient proof of this; and the yard and the measure, the landmark and the coin, show that his ingenuity is made subservient to his sense of justice. By the fine arts he enters the sphere of beauty and the compositions of Mozart and Beethoven, of Handel and Haydn; the pictures of Raphael and Titian, the poetry of Shak[e]speare and Homer must serve to satisfy his longing for intellectual entertainment.

From the above remarks it sufficiently appears, that, though the works of animals are full of intellect and design, it is not *their* design, we admire, but that of the Creator, who accomplishes it through the animals.

RELATION OF INSTINCT TO MAN.

The plastic power, as it exists and operates in the plant is blind, for it has neither sensation nor perception. The instinct of animals is like the twilight, not clear, confused in itself, for it can neither comprehend, judge, nor conclude; it distinguishes only by sensations, and such distinctions are faint unaccompanied by consciousness, and not resulting from comparison, but depending wholly on the strength of the impression made by the different objects on the senses of the animal. The instinct of insects and of those animals, that produce artificial works, is cheerful and so regular in its productions, that they seem to be living arithmeticians; yet they thus only appear to us, for in reality these little creatures are not conscious of their powers. Man has the power of thought; here every thing becomes transparent, clear, distinct and manifest; where pure thought prevails, there instinct loses its power. Though the animal is separated from the vegetable kingdom by sensation and perception, the plastic power of plants is continued in it, and the animal is formed by it, grows, matures and decays according to its laws. So man is separated from the animal by reason, but on the one hand, he is formed by the same plastic power, and on the other, instinct still appears in the new-born child, whose lips long for nourishment, and in many of the actions of savages. The principal manifestations of instinct in man, however, are those of hunger and thirst, of motion and rest, and of care for the race. But, as has been said above, man is separated from the animal by reason; he can think, he can will, and by these powers he reigns over his instinct, and subjects it it to his discipline. He must live, in order to think, he must eat and drink, in order to live; but while the animal is wholly under the control of instinct, and while hunger is a tyrant whose dictates must be obeyed, man can not only eat what he pleases, but also wholly abstain from food, and though urged by an excessive appetite he may nevertheless give away his food, or like Atticus[14] starve himself in the midst of plenty. [. . .]

Man possessing reason, has an innate desire for knowledge, which the animal has not. This desire is more than instinct, and not any part of it. This may be easily made clear. The gratification of instinct is pleasure, connected with sensation and perception, with the taste of food, with the motion of muscles or their rest. But the gratification of a desire for knowledge, is pleasure connected with our apperceptions, with our cognitions, with our comprehensions, judgments, and conclusions. In the sphere of instinct all is confined to sense, in the sphere of our intellectual desires, our pleasure, are derived from thought, from a satisfaction of our thirst for knowledge. But while instinct extends not beyond the sphere of sensation, our desire for knowledge includes *instinct*, and it is this which leads many of us to the objects suited to the exercise of our peculiar talents. So Linnaeus, when yet an infant, could be silenced by no other toy so quickly, as by a flower put into his hands. So Mozart, when only six years old, would make distinctions between different notes, which his father, likewise a musician, could not perceive. Instinct mingling with our desire for knowledge, constitutes in union with it, talents and natural capacities.

The life of man and that of the animals, it must be evident, differ widely, not only in degree, but in kind. Where that of man commences, the animal is left behind. A chasm separates the one from the other. The animal may gaze on man, but it cannot understand him, for it is unable to think or to comprehend, and the words spoken to the dog, for example, are not for him what they are for man, sounds full of meaning, they are mere *signs* to him. The life of the animal is like a dream; and even while waking it dreams. But the life of man is fully awake, it is possessed of self-consciousness and gifted with the capacity to originate constantly new thoughts, and assisted by the powers of nature to realize them.

Editorial Endnotes

1 *Haller's calculation*: Albrecht von Haller (1708–1777), a Swiss anatomist and physiologist known for his studies of the nervous system and muscular contraction.
2 *stylo-pharyngean muscle*: the stylopharyngeus muscle is one of the longitudinal muscles in the throat that assists in swallowing and speaking.
3 *"No bird... winged words fall from the lips of man"*: the phrase "winged words" or variants of it are repeatedly used in Homer's *Iliad* in reference to the speech of various characters.
4 *"morbidezza"*: from Italian *morbido*, meaning "soft" or "smooth." In painting, the term refers to artistic sophistication in creating naturalistic representations of flesh tones.
5 vis sentiens, . . . natura intelligens: this refers, respectively, to a sensitive power and to an "intelligent nature" (and to beings endowed with or embodying either or both).
6 apperceptions: a philosophical (and later psychological) concept referring to the process by which new experiences are assimilated into existing mental frameworks. The concept was developed notably by Gottfried Wilhelm Leibniz (1646–1716) and Immanuel Kant (1724–1804).
7 glebae adscriptus: "registered with/belonging to the soil," denoting a status of feudal serfdom in which labourers were bound to the land and transferred with it in case of a change of landholders.
8 *Reimarus*: Hermann Samuel Reimarus (1694–1768) was a German Enlightenment philosopher known for his criticism of Christianity and his work on the historical Jesus. Rauch's reference is to Reimarus's 1760 book *Allgemeine Betrachtungen über die Triebe der Thiere*, which theorises a system of "drives" or "impulses" that shaped animal modes of life (*Lebensarten*).
9 lotus ornithopedioides: a flowering plant generally found in dry, sandy or grassy places in the Mediterranean Region and Northern Africa.
10 *Descartes*: René Descartes (1596–1650), a French mathematician and philosopher. Descartes argued that animals lacked reason and consciousness and operated solely according to mechanistic principles.
11 *as Aristotle remarks . . . do the same*: Greek philosopher Aristotle (384–322 BCE) discussed the capacities and characteristics of animals in his *History of Animals*, *Movement of Animals*, and other biological writings (to use a modern categorisation).
12 Gaius Plinius Secundus (23–79), commonly known as Pliny the Elder, a Roman philosopher, naturalist, and military commander.
13 Franklin: Benjamin Franklin (1706–1790), printer and publisher (of the important *Pennsylvania Gazette*), scientist and inventor, and one of the founders.
14 *Atticus*: Titus Pomponius Atticus (ca. 110–32 BCE), an author of the late Roman Republic known for his correspondence with the famous statesman and scholar Cicero. After a period of deteriorating health, Atticus decided to expedite things by starving himself to death.

Editorial Headnote

3. Phrenology and Animal "Character" in the Antebellum Period (1845–1851)

3.1 [Orson Fowler], "The Physiology, Phrenology, and Natural History, of the Ourang Outang, or Chimpanze," *American Phrenological Journal* 7, no. 3 (March 1845): 65–72.

3.2 "Animal Phrenology," *American Phrenological Journal* 13, no. 1 (January 1851): 6–7.

3.3 "Animal Phrenology. Number II," *American Phrenological Journal* 13, no. 2 (February 1851): 32.

The beginnings of phrenology, a pseudoscience that would come to enjoy a substantial international career throughout the nineteenth century, can be found in the work of German physician Franz Joseph Gall (1758–1828) and his collaborator Johann Caspar Spurzheim (1776–1832), who later popularised the term phrenology for what Gall himself had termed "organology." Gall's principial dictum was that the brain was the organ of the mind and, following from this, that each of the different mental faculties must be localised within a different part of the brain (or specialised cerebral "organ"). The development of each mental faculty – Gall himself identified a total of 27 – in an individual was indicated by the size of its corresponding organ, while the size of the latter was indicated by the shape of the respective bump on the surface of a person's head. By examining the topography of bumps and depressions, phrenology – or "bumpology," as some of its detractors liked to call it – claimed to allow insights into a given individual's mental and moral traits and tendencies.[1]

Phrenology made its way across the Atlantic in the early decades of the century. Spurzheim himself lectured in America in 1832 (though he never completed his tour, dying of typhoid in the November of that year), as did Scottish lawyer George Combe, another influential proponent and author of the widely read *The Constitution of Man* (1828), between 1838 and 1840. Among the early American phrenologists was physician Charles Caldwell, whose *Elements of Phrenology* (1824) was the earliest American textbook on the subject.

Phrenology's popularity reached its peak around mid-century, though never without the accompanying chorus of criticism and ridicule denouncing it as sham science. From the 1840s onward, the New York City-based publishers and self-taught phrenologists Orson Squire Fowler (1809–1887) and his brother Lorenzo Niles (1811–1896) became the most effective popularisers of phrenology in America and continued to operate as its chief exponents and signal boosters throughout the second half of the century. The Fowlers' widely circulating

American Phrenological Journal – founded in 1838, with Orson Fowler assuming editorship in 1842 – ran until 1911, and those living in or visiting NYC could have a personal look at the skulls and casts on display in the Phrenological Cabinet – located in Lower Manhattan and close to another one of the city's attractions, Phineas Taylor Barnum's American Museum (see P. T. Barnum, Vol. 5) – or even have their own head examined.[2] Soon, phrenology's dubious insights were adopted by a growing cohort of people in various fields, including Denton Offutt, a former general store owner (and early employer of Abraham Lincoln) from Kentucky, who by the mid-1840s had established himself as a successful horse trainer and claimed his training methods to be grounded in phrenological principles.[3]

According to Gall's organology, only a minority of the mental faculties (including those of "religious sentiment" and "poetic talent") were unique to humans, while all others (including pride, affection, and a "sense of numbers") were shared with the rest of animalkind. While "animal phrenology" was not in itself a major focus, the underlying emphasis on human-animal comparability was a basic tenet of phrenological work more generally. As Orson Fowler writes in the first of the following *APJ* articles – a phrenological discussion of "Mademoiselle Fanny," a female orangutan (named after Austrian ballerina Fanny Elssler) whom Barnum had purchased from the Royal Surrey Zoological Gardens for his American Museum – not only did the link between "[o]rganic conditions" and "mental manifestations" extend beyond members of the human species, comparisons between human and animal organisms in fact provided the "strongest argument in proof of phrenological science."

Notes

1 On Gall, see Stanley Finger and Paul Eling, *Franz Joseph Gall: Naturalist of the Mind, Visionary of the Brain* (New York: Oxford University Press, 2019).

2 On phrenology in America, see, for example, Ann Fabian, *The Skull Collectors: Race, Science, and America's Unburied Dead* (Chicago: University of Chicago Press, 2010); Stephen Tomlinson, *Head Masters: Phrenology, Secular Education, and Nineteenth-Century Social Thought* (Tuscaloosa: University of Alabama Press, 2013); Susan Branson, "Phrenology and the Science of Race in Antebellum America," *Early American Studies* 15, no. 1 (2017): 164–93. On the Fowlers and the *American Phrenological Journal*, see James Poskett, *Materials of the Mind: Phrenology, Race, and the Global History of Science, 1815–1920* (Chicago: University of Chicago Press, 2019), chapter 5.

3 See Denton Offutt, *A New and Complete System of Teaching the Horse on Phrenological Principles* (Cincinnati: Appleton's Queen City Press, 1848); Gary A. O'Dell, "Denton Offutt of Kentucky: America's First 'Horse Whisperer'?," *The Register of the Kentucky Historical Society* 108, no. 3 (2010): 173–211. Offutt is credited as one of the first Americans to experiment with a gentler system of horse training that departed from the brutal practice of "breaking." One of his students was a young Ohioan named John Solomon Rarey (see "Educating Animal Workers," Vol. 3), whom he later accused of stealing his methods and betraying his pledge of secrecy. See "The Horse-Tamers in Court. Mr. Offutt's Suit Against Mr. Rarey," *The New York Times*, February 5, 1861, 6.

3

PHRENOLOGY AND ANIMAL "CHARACTER" IN THE ANTEBELLUM PERIOD (1845–1851)

Sources: Phrenology and Animal "Character" in the Antebellum Period (1845–1851)

3.1 [Orson Fowler], "The Physiology, Phrenology, and Natural History, of the Ourang Outang, or Chimpanze," *American Phrenological Journal* 7, no. 3 (March 1845): 65–70.
3.2 "Animal Phrenology," *American Phrenological Journal* 13, no. 1 (January 1851): 6–7.
3.3 "Animal Phrenology. Number II," *American Phrenological Journal* 13, no. 2 (February 1851): 32.

3.1 THE PHYSIOLOGY, PHRENOLOGY, AND NATURAL HISTORY, OF THE OURANG OUTANG, OR CHIMPANZE.

ORGANIC conditions, as influencing and harmonizing with mental manifestations, are not confined to the human species merely. They extend throughout the whole animal kingdom. Indeed, the strongest argument in proof of phrenological science, is that drawn from the organism of the brute creation compared with that of man, and its coincidence with mental manifestation. Nor is the study of the brute creation devoid of interest, as regards the great philosophical *lessons* it teaches, touching the conduct and habits of man. Thus, from the food of different races of animals, may man learn lessons of dietetics as regards his own food. That certain kinds of food are calculated, in and of themselves, to nourish and develop certain portions of the body and certain organs of the mind, and other kinds, other portions and organs, is a principle of philosophy and physiology abundantly attested by the strongest proof, as well as important beyond all expression or conception; for, by applying it properly, we can *feed,* through the alimentary canal, the muscular system, when it is too feeble, or the mental, or the vital, or any organ of the vital, as occasion may require, and also particular *ranges* of the mental organs, such as the animal, the moral, or the intellectual, and even *particular organs,* Thus, analogy renders it *certain* that animals have aptitudes or relishes for those kinds of food most, that best nourish those physical and mental powers which constitute the nature of the animal in question. He who questions this *summary* of analogical reasoning, must deny *all* analogy – all induction. Now, the tiger craves *animal* food, still warm with recently extinguished life. Who can doubt but that

DOI: 10.4324/9781003032717-4

this diet is calculated, *by constitution,* to feed both the mental and physical *nature* of the tiger? What, then, is that nature? Why, muscular power and mental fierceness and cunning. In other words, the fact that the tiger constitutionally craves animal food, shows that a flesh diet is naturally calculated to feed *Destructiveness and Secretiveness;* for, if it is constitutionally adapted to feed the nature of the tiger, which consists mainly of these organs, it is equally calculated to feed these same organs in man. The same principle applies to nuts, as being calculated to develop the nature of the squirrel, and other kinds of food as being adapted to the constitutional peculiarities of other animals. This principle furnishes man with a dietetic guide, for developing whatever part or parts of his animal or mental economy may be required to be developed.

Other great truths, as adapted to the study of human character, as drawn from the general physiognomies of different animals, and as adapted to the promotion of the great ends of our animal and mental economy, may be drawn from the study of animals, their physiological and other habits, their adaptation to nature – being taught us by the constitutions of different species of animals. But these lessons have never been read by man. Animals have been studied *independently.* They should be studied with reference to their relations to the great *whole* of nature. And it is devoutly to be hoped, that *phrenological* naturalists will soon come forward, who shall treat this subject on the extended scale here pointed out.

[. . .]

A few days ago, the proprietor of Miss Fanny, the female Ourang Outang now exhibiting in the American Museum, invited the Editor to make a phrenological examination of this half human subject. Handing a brush, (for she is very fond of being *combed,*) I applied it with one hand, while the other was passed over her head, by way of ascertaining her phrenological organization; this affording a most excellent opportunity of observing them correctly.

They are as follows: – The size of her brain, as a whole, may, perhaps, have exceeded that of an infant at birth, though I should question even this. What brain there is, is located almost exclusively in its base, as will be seen from the annexed engraving, which is substantially correct, as far as the shape of the head is concerned.

Her body is less full than is here represented, though this may have been occasioned by its having grown poor from a change of habits, consequent on its introduction into civic life; this species of animal generally living but a short time after this change.

Her head rises about two and a half, perhaps three, inches above the opening of the ears, but is conical, or short and narrow, on the top. The organs mainly developed, are, the social, the animal, and the perceptives. A slight development of Causality is discernible, though but very slightly so. Comparison is plainly perceptable, the head rising at the upper part of the organ about an inch and a quarter above the orbits of the eyes. Benevolence is also quite apparent, the head rising higher at this point than at any other, except at Firmness, where it forms a kind of apex. Between Benevolence and Firmness, or at Veneration, the head falls

NO. 6. – THE OURANG OUTANG.

in very considerable, though this organ is not wholly wanting. I should be loath to declare, phrenologically, that she is *wholly* incapable of religious feelings, yet she may be.

Directly from the two sides of the upper part of Benevolence, two ridges, projecting above the other organs, as plainly as would one's finger if laid upon a flat surface, standing out thus prominently, are the organs of Imitation. For the *qualities* imparted by this faculty, the whole monkey race are proverbial. I pointed out to her owner these conspicuous ridges. He expressed his surprise that protuberances as great as these, had hitherto escaped his observation; and then related instances in illustration of her imitative power and disposition. Conscientiousness is but slightly, if at all, perceptible. Firmness is conspicuous, and so is Self-esteem, the largest organs, with two or three exceptions, in her head; and, during my short stay, she abundantly evinced the possession of these mental qualities. When I first entered the apartment she occupied, she was trying to open a trunk with a bunch of keys that they had handed to her for the purpose of selecting and using the one that opened it. Several times, she was called off to allow my observation of her head, but she pertinaciously refused to give up the task she had undertaken. Continuity (Concentrativeness,) was fairly developed, though not enough so to fill up the place of its location even with Self-esteem above, or the social organs below. Hence, it presented a partial cavity. In the succeeding account of the unwillingness of these animals to be captured, abundant evidence will be seen of their possessing both Firmness and Self-esteem, or at least love of liberty. Approbativeness is almost completely wanting. At least, I could discover little if any, traces of its development. Cautiousness is discernible, yet it is a small organ, as will be seen in the engraving, from the narrowness and rapid sloping of the head from Firmness to Destructiveness. This also indicates the comparative absence of Conscientiousness. The social affections are all strong. Amativeness is apparent, but not by any means predominant. It will, however, be remembered, that she was only four years old – too young for its relative size to be as yet determined with certainty. Love of young was the predominant organ in this group. Adhesiveness was plainly observable, yet not predominant. This probably accounts for the generally observed sadness and loneliness of the animal after its capture, and coincides with the *gregarious* habits of the species.

Combativeness, Destructiveness, and Appetite, were her largest organs. The former presented great ridges, larger, even, than those of Imitation; and she evinced the quality in question most forcibly. Thus: a tobacco-box was given her, which, by accident, she dropped. A boy standing by, picked it up, evidently with a view of giving it back to her; but, before he had time to do so, as he was raising himself up, she grappled him by the cap and hair together, and yerking off the latter, slung it down, and was proceeding to other demonstrations of violence, when she was pacified by the re-delivery of the article she fought for so bravely. During my short stay, some other offence, I forget precisely what, was offered, and she chattered away most combatively at the offender. I say combatively, because, by this, reference is had to the *tones and natural language* of this faculty; for language consists, not merely, not mainly, in the use of words, but in all that gesticulation by which we communicate ideas or feelings. Her dialect, I could not understand, but her manner and enunciation I did comprehend very distinctly. Would that I could give the reader an idea of it. It consisted, as near as I could judge, of but a single word, something like our words, chat or not. There might have been two or three, nearly alike, but she seemed to me just as if she *were forbidding* the thing in question. I could not help thinking of an old woman, with a small brain, but all flustered with rage, scolding in Hebrew or Hindoo. From hearing its intonation, and listening to its short, sharp enunciation and harsh voice, I doubt whether any one, even without *seeing* what she *acted* as well as spoke, would have been at a loss as to her expressing anger, and that as plainly and as forcibly as we do to each other by *word* as well as by manner. From what I saw, I was lead to the belief, that this species of animals is not wholly incapable of learning to talk, and that they have a rude dialect of their own. The keeper says, he has taught Ourang Outangs to say yes, and to use the word correctly whenever it was required to express assent.

Destructiveness is also large, as will be seen faithfully represented in the preceding cut. She, however, gave no particular indication of it, except those above enumerated in conjunction with Combativeness. But her manner betokens it, especially the great *force* she throws into what she does. Secretiveness is fully developed, yet it is by no means equal to either Combativeness or Destructiveness. Alimentiveness is largely developed, and Acquisitiveness is quite conspicuous. She eats heartily, and evinced the feeling of *it is mine,* quite forcibly in the matter of the tobacco-box above alluded to. Ideality is almost wholly wanting, as is Spirituality, (commonly called Marvellousness,) and Mirthfulness; but there is a slight development of Constructivenees, though but slight, and she manifests this in sewing. She has been known to laugh.

Of the development of her perceptive organs, the reader can form a tolerable idea from the annexed cut. A difference will be observed in the position of her eyes, compared with those of the brute creation generally. In general, the eyes are in the *sides* of the heads of animals. In no instance, except in the human species and the monkey races, do the perceptive organs extend *over* the eyes sufficiently to form an arch or a roof over them, and nearly encircle them. But in these, they do. This indicates a far higher development of the perceptive *organs,* which harmonizes with their possession of these phrenological *faculties* in much greater power and activity, than the balance of the brute creation. [. . .] To what extent they evince order, I have at present no knowledge. Fanny's keeper mentioned that she showed a slight capacity of numeration, in *missing* things where some of a number had been removed. Her Individuality is large in the head, and appeared abundantly developed in character, both in the quick, keen, and almost constant roll of her piercing eye, and in the fact, that she observed things so quickly and so instantly. Form is also large in the head, and also in character. Instances of her remarkable recognition of persons, were narrated. She generally gave her hand to those introduced to her.

Taking her as a sample of her race, both in character and development, it will be interesting and instructive to compare what is known of the character, disposition, and habits, of her species, with this account of her developments. Perhaps, however, it may not be unimportant to add here, that, in 1835, I saw a rather larger and more powerful animal of this species, in Philadelphia, the developments of which much resembled those of this sample, except that its head was relatively wider and flatter, and its character more fierce, the one in the museum being remarkably domesticated, a quality certainly most desirable, both to the casual examiner, and to the scientific observer. I saw another in Troy, in 1834, between the baboon and the orang outang, a most fierce and powerful animal, and completely unmanageable from its rabid Destructiveness, probably inflamed by its captivity. It did not survive long, its restlessness, its treatment, and our northern climate, conspiring to produce premature death.

3.2 ANIMAL PHRENOLOGY.

THE HORSE, NO. 1.

The horse is the noblest and one of the most intelligent of the lower animals. Like the dog, he seems fitted to be the companion as well as the servant of man. As men are unlike in disposition, the same is equally true of horses. While one exhibits great pride, another is dull, tame, and humble – one is kind, another savage – one is bright and intelligent, another stupid, and comparatively unteachable – one is courageous, another is timid and shy, and therefore dangerous to drive. And these differences run through all the faculties common to the species. Nor is it true of the horse only, but is equally applicable to all animals.

Phrenology points out many of these differences, and is an important aid to those who deal in, or train and use horses. If we can learn to detect at a glance, by the shape of the head, the vicious, intractable horse, or the mild, courageous, intelligent, teachable one, this single feature of knowledge would compensate for all the trouble and expense of learning the science.

Width between, and prominence of the eye, indicate intelligence, the faculty to learn and understand our wants, and the adaptation to learn to work, and perform feats, tricks, and the like. All learned horses in the circus are of this description.

Roundness and elevation between and above the eyes indicate mildness and amiability, and a desire to be caressed and to reciprocate kindness.

Width between the ears indicates courage, nobleness, and strength of character, patience and energy.

A timid, skittish horse, is narrow between the ears like the deer, sheep, and rabbit, showing small Destructiveness and Combativeness.

A dull, weak minded, and unteachable horse, is narrow between the eyes, which are not prominent, and flat and contracted above and back of them.

Temperament, of course, is just as influential in the horse as in man, and can be understood with little trouble. We may give an article on this subject in a future number.

In these heads a marked difference will be seen, No. 1, is much broader between the eyes, more full, rounded, and prominent in the forehead, broader between the ears, and much more full between the eyes and ears, and altogether has a larger brain. This horse can be trusted, and he loves and trusts man. He will stand anywhere, will be animated, not frightened, by noise or show, like military parade and music; will draw or hold back well, the harness breaking and the vehicle hitting his heels will not frighten or make him kick; he can be used or played with by women and children, and is a first-rate trust-worthy, family horse. Moreover he can easily be taught almost anything, and is sufficiently spirited without being fiery or cross. The large nostril also shows good wind and endurance.

NO. 2.

No. 2, shows a marked contrast with the other in almost every respect. His intelligence is confined merely to self-preservation, and he is destitute of kindness, mildness, and tractability: will bite, kick, or run away at every opportunity, or balk if he is irritated, or has a harness-chafe: is easily frightened at empty barrels or anything unusual, yet, whatever he can drive, or frighten and bite, as, for example, a boy, a sheep, or a goose, he is forever tormenting. He lacks every element of nobleness, elevation, and amiability, and is only fit for a mill or horse-boat. For general use he should be avoided.

The Dog.

NEWFOUNDLAND DOG.

The fact that dogs are capable of a high degree of development is well known throughout the civilized world, yet the marked difference in their PHRENOLOGY may not be so generally apparent. The accompanying illustrations exhibit the difference between the common Newfoundland dog and the wild, uncultivated dog of the forest and the prairie.

All varieties of the dog are capable of culture, yet there is as much difference in their tractability as there is between Anglo-Saxons and native Indians, and the civilization of dogs as well as of men, changes their organization.

In the domssticated [sic] dog we find a high development of the INTELLECT and of BENEVOLENCE, while in the wild dog the absence of both these qualities are, to a phrenologist, at once perceptible.

WILD DOG.

How much higher is the development, of the head of the Newfoundland above and back from the eyes than in the wild dog. In one the organs of intelligence and kindness are very large, and indicate the highest order of mental capacity and susceptibility to education and the friendship of man. The lap-dog is another instance of a similar development, with larger Combativeness, hence they are more snappish than the Newfoundland. Domestic training and familiarity with man has elevated the character of those animals, and with it their heads.

The wild dog has a broad, low head, flattened on the top, showing very large Destructiveness, Combativeness, and other organs indicating fierce propensities, with very small Benevolence and contracted intellectual organs, making him unfriendly, and ferocious.

3.3 ANIMAL PHRENOLOGY.

NUMBER II.

THE LION.

The kingly prëeminence of this monarch of the forest is indicated by surpassing muscular power, magnanimous courage, and nobleness. His figure is striking, his look confident and bold, his gait proud, and his voice terrible. Lavater[1] says of him: – "What majesty in the countenance of the lion! How far from mean, insidious cunning, ensnaring ferocity. It is a ferocity of a different kind, of conscious strength and superiority. *In the region above the eyes* appear *consideration* and *discretion.*" The elevation and arching of the forehead indicate intelligence and magnanimity which serve to ennoble the ferocity of the lion. He will not worry or destroy small animals, as will the tiger, fox, and cat, but seems to be governed by a feeling of honor. His ferocity is never malignantly exemplified as if to torture the inferior, it is in self defense and for food that the tremendous energy of the lion is awakened. By his superior intelligence, he is susceptible of a high degree of cultivation. Taken young, he fraternizes readily with man, and becomes very fond of him. His consciousness of power, and his native pride of character teach him not to permit rough treatment, and man becoming afraid of that power, ultimately loses control over him. Were man larger and stronger than the lion as he is superior to the dog, then the lion would be alike his pet, friend, and servant. The lion often permits the little jackal and other small animals to live in peace with him, and avails himself of their skill in taking game. The tiger is more cruel, cold, unsocial and selfish, and instead of protecting the jackal from harm, would destroy and make a supper of him the first time a supply of other game should fail.

The lion is as much superior in character to the other carnivorous animals of nearly his own size, indeed, of the feline race generally, as the Newfoundland and Mastiff dogs are more noble than the little, snappish, whiffet.

The lion is said to possess, even in his wild state, an appreciation of man's superior wisdom, and the sentiment of gratitude for kindnesses shown him when in distress. It is related that a lion having a thorn in his foot, carefully approached a person, walking on three legs holding np the lame one, and apparently imploring aid. The person took the hint, drew out the thorn, and the lion gave him such a look of gratitude as he returned to his lair, as to immortalize his name for forbearance, nobility, and intelligence. This same lion is also said to have afterwards been captured and placed under the control of a king, in a cage, or "lion's den." It so happened that this man fell under the displeasure of the king, and was condensed to be cast into the den to be devoured by the lion. The moment he reached the ground, the lion came rushing to seize him, but stopped suddenly on perceiving that it was his old beneafctor [sic] who had years before extracted the thorn from his foot, when both parties were free, and he began caressing him like a favorite dog. This fact so wrought upon the superstitious feelings of the king, that he pardoned the man, and gave him the lion as a companion, who followed him wherever he went with perfect docility and enduring affection.

THE FOX.

The fox is the embodiment of selfishness and treachery. Cunning is his leading trait, and all history gives him a bad name. He is sly, artful and intriguing, not only in his efforts to secure game for himself, but to avoid being made game of by others. He cannot be accused of mental dulness, or intellectual inferiority, but he has so little pride and nobleness, and so much finesse, that he degrades himself by acts of meanness, which a lion or a respectable dog would despise. It is very difficult to set a trap which he will not detect and avoid, and when chased by dogs, he is seen to run on ice, in the water, on a log, and often cross his track, and frequently run in a straight line, and suddenly leap off at one side and take the opposite direction, all apparently to bewilder the dogs, and thereby facilitate his escape.

Nor is the fox in much hurry when chased, appearing to depend upon skillful tricks and gyrations, to hinder the dog and make him run two or three times as far as he runs himself; and it is seldom that he will run out of the hearing of the dog, but stop and wait for him. He will often return to an old track and run in the opposite direction to mislead his honest-hearted pursuer.

We knew a young domesticated fox who would gambol during the day among the children and chickens as honestly as a respectable dog, but, although well fed on fresh meat, he would steal out of his kennel, when all was still, and kill several chickens, and bury them in the garden, and having eaten one or two, would return to his kennel and look as honest in the morning as if nothing had happened.

Secretiveness and Cautiousness are very large in the fox, as seen in the great width of his head, and his cunning expression of face; that shrewd smirk, as seen by the position of the eyes and sharp phiz, harmonize with his head and character. We have never seen the skull of any animal which equals that of the fox for its development of Secretiveness. The following, related by an English paper, illustrates this remarkable trait of the fox:

> A farmer had discovered that a fox came along a beam in the night to seize his poultry. He accordingly sawed the end of the beam nearly through. In the night the fox fell into a place whence he could not escape. On going to him in the morning, he found him stiff, and, as he thought,

lifeless. Taking him out of the building, he threw him on the dung hill, but in a short time Reynard[2] opened his eyes, and seeing all was safe and clear, galloped away to the mountains, showing more cunning than the man who ensnared him.

Editorial Endnotes

1 *Lavater*: Johann Kaspar Lavater (1741–1801) was a Swiss philosopher and theologian and a populariser of (the modern variant of) physiognomy, which held that personality traits could be determined from an individual's facial features and other aspects of physical appearance.

2 *Reynard*: Reynard the Fox is a folkloric anthropomorphic animal figure – generally represented as a cunning and crafty trickster – that first appeared in the twelfth century in a number of medieval French, Dutch, and German fables.

Editorial Headnote

4. Lewis Henry Morgan, *The American Beaver and His Works*. Philadelphia: J. B. Lippincott and Co., 1868.

Pioneering anthropologist Lewis Henry Morgan (1818–1881) was one of the more prominent nineteenth-century figures guilty of the "monstrous heresy" of assuming that "the whole animal kingdom, including man, has 'one psychical nature.'"[1] Published between his early ethnographical studies of the Iroquois and his well-known later work on systems of kinship and social evolution, *The American Beaver and His Works* is an elaborate study of the industrious "mutes" that also articulates a rebuttal of contemporary arguments against animal intelligence. In *The American Beaver*, Morgan builds on several points already presented in a two-part article on "Mind or Instinct," published pseudonymously in 1843, which criticised the tendency to subsume all forms of seemingly intelligent animal behavior under the rubrum of instinct. For Morgan, the concept of instinct was obfuscating because what it referred to was nothing other than mind, and if human beings were identified as the authors of the very same actions, "we should without hesitation pronounce [them] the consequence of abstract consideration" instead of relegating them to the mysterious workings of instinct.[2] Instinct was "a designation that prohibits inquiry, because it pretends to furnish an explanation of itself" – it was a strategic concept whose function consisted in bolstering claims of human uniqueness.[3] As he puts it in his chapter on "Animal Psychology" in *The American Beaver*, from which the following excerpts are taken, instinct was "an invention of the metaphysicians to assert and maintain a fundamental distinction between the mental principle of the human species and that of the inferior animals."[4]

Notes

1 "Natural History of Man," *The United States Democratic Review* 27, no. 145 (July 1850): 41–49, here 42.

2 Lewis Henry Morgan [as "Aquarius"], "Mind or Instinct: An Inquiry Concerning the Manifestation of Mind by the Lower Orders of Animals," *The Knickerbocker* 22, no. 5 (November 1843): 414–20, here 418.

3 Lewis Henry Morgan [as "Aquarius"], "Mind or Instinct: An Inquiry Concerning the Manifestation of Mind by the Lower Orders of Animals (Continued)," *The Knickerbocker* 22, no. 6 (December 1843): 507–15, here 514.

4 On Morgan's beavers, see Marc Swetlitz, "The Minds of Beavers and the Minds of Humans: Natural Suggestion, Natural Selection, and Experiment in the Work of Lewis Henry Morgan," in *Bones, Bodies, Behavior: Essays on Biological Anthropology*, ed. George W. Stocking (Madison: University of Wisconsin Press, 1988), 56–83; Gillian

Feeley-Harnik, "The Ethnography of Creation: Lewis Henry Morgan and the American Beaver," in *Relative Values: Reconfiguring Kinship Studies*, ed. Sarah Franklin and Susan McKinnon (Durham: Duke University Press, 2001), 54–84; biographical information are taken from Thomas R. Trautmann, "Morgan, Lewis Henry," in *American National Biography*, ed. John A. Garraty and Mark C. Carnes, vol. 15 (New York: Oxford University Press, 1999), 848–51.

4

ANIMAL PSYCHOLOGY (1868)

Lewis H. Morgan

Source: Lewis Henry Morgan, *The American Beaver and His Works* (Philadelphia: J. B. Lippincott and Co., 1868).

THE popular mind has always been in advance of the metaphysicians with reference to the mental endowments of animals. For some reason there has been a perpetual hesitation among many of the latter to recognize, in the manifestations of the animal mind, the same characteristics that are displayed by the human intellect: lest the high position of man should be shaken or impaired. Besides this, the connection in man between the intellectual faculties and the moral sense is found to be so intimate, that the concession of the former has seemed, to cautious minds, to draw after it the necessary admission of the latter. In attempting to escape this imaginary dilemma, the metaphysicians have been betrayed, as it would seem, into a false position. This is shown by the invention, in modern times, of a vague, not to say fictitious, principle, with which all animals have been arbitrarily endowed for the government and maintenance of their lives. There can be no objection to the use of this principle, which is termed "instinct," to explain, or rather to leave unexplained, certain mental phenomena exhibited equally by mankind and the inferior animals, so long as it is restricted to those mental processes which are beyond the reach of consciousness. But the attempt to explain all the mental phenomena manifested by the mutes by means of an arbitrary term is an evasion of the true question involved. It would be difficult, in right reason, to discover the slightest tendency to lower the personal dignity of man, or to alter in the least his responsibility to God, by recognizing the existence in the mutes of a thinking self-conscious principle, the same in kind that man possesses, but feebler in degree; nor even by conceding their possession of a moral sense, although, so far as our present knowledge extends, it is so faintly developed as scarcely to deserve the name. Man, at least, should neither admit nor deny the moral sense to the lower animals because of the supposed bearing of such an admission upon his own relations to the Supreme Being. The question of the degree and kind of their mental endowments should stand upon its own basis, and be resolved upon its own merits. I trust the sensibilities of no one will be disturbed by this method of introducing the subject of Animal Psychology; and that the subject may be considered unaffected by external complications, and be studied independently upon its own authoritative facts.

DOI: 10.4324/9781003032717-5

When the Creator brought into existence the various species of animals, He intrusted to each individual being the care of his own life. As a principle of intelligence was indispensable to capacitate each one to maintain and preserve that life, we find each individual endowed with a mental or spiritual essence which is distinct from the body, but associated with it in a mysterious manner. It requires no argument to prove that the mutes possess a principle of intelligence which performs for them the same office in governing their conduct that the human mind does for man. When the existence of mind in the mutes is recognized, the qualities it manifests become the subject of investigation. As we know nothing of the ultimate nature of the human mind, so in like manner we know nothing of the ultimate nature of the animal mind; but since the former manifests certain faculties, as memory, certain passions, as anger, certain appetites, as hunger, and puts forth a certain power, the will, – the true inquiry is, whether the latter manifests certain faculties, as memory, certain passions, as anger, certain appetites, as hunger, and puts forth a certain power, the will? If the affirmative is found to be true as to each of these propositions, then the next question must be, whether any difference in kind can be discovered between the memory of a man and the memory of a mute; between the anger of the one and the anger of the other; the hunger of the one and the hunger of the other; or the will of the one and the will of the other. Unless some real and determinate difference can be found by which to differentiate the qualities of the animal mind from those of the human mind, it must necessarily follow that the mute and the man are both endowed with a similar mental principle; and that man owes his superior dignity not to the exclusive possession of this principle, but rather to its enjoyment in a higher, more ample, and more distinguishing degree.

It is one of the extraordinary features of this Divine gift that it is capable of adaptation to so many, and to such diversified organisms; and not less remarkable that it should still reveal the fundamental similarities of a common principle through all its ramifications, so far as we are able to observe its manifestations. Our knowledge of the lives of the higher animals is extremely limited, and founded upon observation alone; while of the inferior species it is next to nothing. The discussion of the subject of Animal Psychology is, therefore, necessarily limited to the higher animals, and to such facts, with reference to these, as are well authenticated and universally admitted. Any argument which passes beyond the range of ascertained facts is incapable of proving or disproving any proposition.

Neither is it desirable to perplex ourselves with the question, whether or not the mutes possess a conscience, or the moral sense. While a negative declaration proves nothing, an affirmative assertion is without support in existing knowledge. The prior question, in point of time, is concerning their mental endowments.

It is equally unnecessary to discuss the grounds of the artificial distinction which is made between the appetites and passions on the one hand, and the intellectual powers on the other. The concession of the former to the mutes in common with mankind, and the withholding of the latter as an independent and distinguishing gift, is an assumption which tends to mislead without advancing the true inquiry. The passion of anger and the pain of hunger can only be predicated of a mental principle, of which they are manifestations as absolutely as memory or imagination. Indeed, it is an axiom in moral as well as in intellectual science, that pain and pleasure are experienced in the mind, and not in the organs of the body. When, therefore, we find the phenomena of pleasure and pain displayed by individuals of every species, and to be essentially the same in kind among them all, it leads to the same general conclusion; namely, that all living creatures possess a similar mental principle. This leaves the question of difference in degree, which was rendered necessary by difference in species.

I propose to submit, in a brief form, a series of considerations or arguments based upon the structural organization, and authenticated acts, of the higher animals, tending to show: first, that they possess a mental principle; secondly, that the qualities which it manifests are essentially the same as those displayed by the human mind; and lastly, that the difference between these qualities, and, inferentially, between the principles they respectively represent, is one of degree and not of kind. The discussion, to be brief, must necessarily be general; and it is entered upon rather for the purpose of offering suggestions upon branches of the subject, than of treating it systematically as a whole. I have neither the facts nor the ability to prepare a treatise upon this important but difficult theme.

I. *Structural Organization.* It has been demonstrated, by anatomical comparisons, that the structural organization of the vertebrate animals conforms to a general plan, the fundamental features of which run through all the species, genera, orders, and classes of this branch of the animal kingdom. The several species thus stand in fixed relations to each other, and are all bound together by the common creative thought which is incorporated in the diversified forms of the individual representatives of each. Man, therefore, is not permitted to overlook the fact that he is a constituent member of this vertebrate branch; and although endowed, relatively, with the highest capacities, and invested with the highest organization, he cannot free himself from the bond by which its several members are indissolubly united.

Among the conspicuous features of this plan of structure is the brain, which is enveloped in a skull, and placed in immediate connection with the organs of sense. The nervous system, of which the brain is the centre, is universally regarded as the seat of the mental principle. Since all the vertebrate animals possess both the one and the other, they are all alike raised to the first condition necessary for the manifestation of intelligence. In the next place, they all agree in the possession of the organ of vision, located in the head in immediate connection with the brain; of the organs for smelling and hearing (with the exception perhaps of some species), similarly placed, and holding similar relations to the brain. Besides these, are the senses of taste and touch. These several senses, operating through similar mechanisms, have but one office, that of communicating impressions of external objects to the brain for the information of the mental principle. By their means a second condition of intelligence is secured; namely, perception. Without one or more of these senses, which are the instruments of perception, the bare continuance of animal life would be impossible; and yet, without the presence of a mental principle to take cognizance of the impressions thus conveyed, their object would necessarily be defeated.

These intimacies of structure are particularly remarkable among the mammals. The office and functions of the several bones and muscles of the animal frame are much the same in the different species. So the nervous system, which is centralized in the brain, is distributed throughout the body in such a manner that the relative position as well as functions of its several parts are similar, if not precisely the same, in all. The several ganglia are found in the same connection with the nerves of sensation and of motion, and performing the same offices in a similar manner. Such minute differences as exist find their explanation in the special adaptation of each animal to his sphere of life. In like manner, the circulating system is constructed upon the same general plan, employing the same organs, with slight variations of form. The same is equally true of the organs of respiration and of the digestive apparatus. One nomenclature suffices for the minutest subdivisions of the mammalian form.

[. . .]

The structure of the higher animals leads directly to the inference that each of their organic forms was designed to be actuated and governed by a thinking principle; a principle not only capable of receiving impressions conveyed by the organs of sense, but also of making a rational use of the perceptions which these organs were designed to throw perpetually under its cognizance. To deny the existence of the principle, or its power to act, is a denial of the obvious purpose of the elaborate mechanism of the animal frame.

From every point in which the structural relations of the vertebrate animals are considered, a common plan of creation is not only seen, but this, in turn, becomes deeply significant upon the question of similar mental endowments. These intimacies of structure are the foundation of corresponding intimacies in the principle of intelligence by which they are actuated.

II. *The Principle of Life.* Life in all its forms is a mystery. As a formative power, it builds up the infantile body from weakness into maturity and strength. It maintains a perpetual conflict with the elements of disorder and decay until the organism in which it dwells breaks up, or wears itself out. Is death the destruction of this principle? or is it immaterial, and expelled, like the spirit, from the body? If it be a principle, and, therefore, immaterial, it would be difficult to show that the living and thinking principles are separate and distinct entities. It seems to be more than surmisable that the two are identical. It is I – the spirit – which lives, and not the body, which is material. If life comes of the union of body and spirit, then it is not an entity, but a result; and all there is of life is the life of the spiritual essence, or of the principle of intelligence.

Vegetable life cannot be compared with animal, because the former, to omit other differences, is without self-consciousness. Will it be said that the mutes are without consciousness? It is answered that consciousness is an inseparable and essential quality of the mental principle. When a beaver stands for a moment and looks upon his work, evidently to see whether it is right, and whether anything else is needed, he shows himself capable of holding his thoughts before his beaver mind; in other words, he is conscious of his own mental processes.

The possession of the principle of life by the higher animals, from its most robust to its most sensitive forms, draws after it whatever this principle may represent.

III. *Memory.* The mind is known by its qualities exclusively.

As a principle, or essence, it is not divisible into parts, or faculties, or organs, each having an independent existence. "The utmost ingenuity," says Abercrombie,[1] "has not been able to advance a step beyond the fact that the mind remembers, reasons, imagines; and there we must rest contented."

It cannot for a moment be doubted that the animal mind remembers, and that it displays this quality as purely and as absolutely as the human mind. Memory, then, must be conceded to be one of its qualities. Its quickness or slowness, its retentiveness or weakness, are wholly immaterial. It is sufficient that the animal mind is able to recall a former perception, or previously known fact, and to have treasured it during the interval. The inference that follows from the recognized possession of a principle capable of remembering is very important. Memory is one of the qualities by which the existence of the human mind is demonstrated. By the same quality the existence of a corresponding principle in the mutes is also established. If a comparison of the two acts of remembrance show them to be in all respects similar, then the two principles of which they are manifestations are, inferentially, the same in kind. The difference is indeed immense between the memory of a familiar object, or even of a series of antecedent facts, which a mute may exhibit, and that powerful memory in man, which not only is able to hold the facts of universal knowledge, but also to reproduce the process of reasoning, by which the great truths of science have been demonstrated. This difference, however, is immaterial, since it is one of degree, and not of kind. As there is a gradation of its power among the individuals of the human species, so there is undoubtedly a similar gradation among the several species of the vertebrate animals.

"Memory," says Sir William Hamilton,[2] "is an immediate knowledge of a present thought, involving an absolute belief that this thought represents another act of knowledge that has been." As the mind is a unit, the whole mind remembers, and not one of its fractional parts. If the power to remember were removed from the mental principle, it would become powerless, and perhaps be overthrown. The past, in such a case, would be utterly lost, the present vanishing with every instant, the future inconceivable, and the external world a blank. On the other hand, let any created being possess, in addition to the senses, a something capable of remembering, and it has more than the power to remember; it has, with it, a capacity to know, to understand, and to reason. That something is the mental principle. Every other inference is excluded. Knowing the qualities of this principle as it exists in the human species, and conscious of its unique and extraordinary character, when we find the mutes in possession of a something which displays the same qualities, the philosophical axiom at once suggests itself, namely, "that a plurality of principles is not to be assumed, when the phenomena can possibly be explained by one."[3]

IV. *Reason or Judgment.* The mutes perceive external objects in the same manner that we do. After admitting that no distinction can be found between their manner and our own of acquiring a knowledge of external things, through the organs of sense, it has been denied that they are able to make a rational use of the perceptions thus obtained. Their acts, in innumerable instances, are seen to be acts of intelligence and knowledge, such as a man would perform under similar circumstances, and yet, there is an unwillingness to recognize in them the results of deliberate processes of reasoning, followed by an exercise of the will. A large class, it is true, acknowledge some reasoning powers in the mutes, but under such qualifications, limitations, and restrictions, that, in effect, it denies to them the possession of a free intelligence. The real question is practically evaded. Their acts should be tested by the same analysis which is applied to human acts, and full credence be given to the results. As we cannot place ourselves in personal connection with the animal mind and thus obtain their testimony concerning their mental processes, we are remitted to their personal acts. Upon these, however, a judgment can be formed as definitely as one man can pronounce upon the act of another man. While this method is not as irrefragable as an appeal to consciousness, it is one upon which mankind act implicitly in their own affairs.

[. . .]

The works of the beaver afford many interesting illustrations of his intelligence and reasoning capacity. Felling a tree to reach its branches involves a series of considerations of a striking character. A beaver seeing a birch-tree full of spreading branches, which to his longing eyes seemed quite desirable, may be supposed to say within himself: "If I cut this tree through with my teeth it will fall, and then I can secure its limbs for my winter subsistence." But it is necessary that he should carry his thinking beyond this stage, and ascertain whether it is sufficiently near to his pond, or to some canal connected therewith, to enable him to transport the limbs, when cut into lengths, to the vicinity of his lodge. A failure to cover these contingencies would involve him in a loss of his labor. The several acts here described have been performed by beavers over and over again. They involve as well as prove a series of reasoning processes undistinguishable from similar processes of reasoning performed by the human mind.

Again, the construction of a canal from the pond across the lowlands to the rising ground, upon which the hard wood is found, to provide a way for the transportation of this wood by water, is another remarkable act of animal intelligence. A canal is not absolutely necessary to beavers any more than such a work is to mankind; but it comes to both alike as the result of progress in knowledge. A beaver canal could only be conceived by a lengthy and even complicated process of reasoning. After the conception had been developed and executed in one place, the selection of a line for a canal in another would involve several distinct considerations, such as the character of the ground to be excavated, its surface, elevation above the level of the pond, and the supply of hard wood near its necessary terminus. These, together with many other elements of fitness, must be ascertained to concur before the work could be safely entered upon. When a comparison of a large number of these beaver canals has demonstrated that they were skillfully and judiciously located, the inference seems to be unavoidable that the advantages named were previously ascertained. This would require an exercise of reason in the ordinary acceptation of the term.

And this leads to another suggestion. Upon the Upper Missouri these canals are impossible, from the height of the river banks; and besides this they are unnecessary, as the cotton-wood, which is the prevailing tree, is found to the edge of the river. While, therefore, canals are unknown to the Missouri beavers, they are constantly in use among the beavers of Lake Superior. On the other hand, the "beaver-slides" so common and so necessary on the Upper Missouri, are unnecessary, and therefore unknown, in the Lake Superior region. Contrary to the common opinion, is there not some evidence of a progress in knowledge to be found in the beaver canal and the beaver-slide? There was a time, undoubtedly, when the canal first came into use, and a time, consequently, when it was entirely unknown. Its first introduction was an act of progress from a lower to a higher artificial state of life. The use of the slide tends to show the possession of a free intelligence, by means of which they are enabled to adapt themselves to the circumstances by which they are surrounded. In like manner it has been seen that the lodge is not constructed upon an invariably typical plan, but adapted to the particular location in which it is placed. The lake, the island, and the bank lodge are all different from each other, and the difference consists in changes of form to meet the exigencies of the situation. These several artificial works show a capacity in the beaver to adapt his constructions to the particular conditions in which he finds himself placed. Whether or not they evince progress in knowledge, they at least show that the beaver follows, in these respects, the suggestions of a free intelligence.

"Instinct," says Dr. Reid,[4] "is the habitual power of producing effects like contrivances of reason, yet so far beyond the intelligence and experience of the agent, as to be wholly unexplainable by reference to them." Habitual acts can only be understood from human experience. Acts to be performed habitually or mechanically must first be learned by an exercise of intelligence. It is a very unsatisfactory explanation of the works of a beaver, to affirm that he was endowed at his birth with a mechanical skill which, by the laws of mind, must be acquired by experience. An assertion that the acts of a beaver in felling a tree, in constructing a dam, or in excavating a canal, are beyond his intelligence, is mere assumption, as well as a contradiction of terms. This conclusion flows legitimately from the original blunder of attempting arbitrarily to endow animals with a supernatural principle, which enables them to perform ignorantly and blindly works of intelligence and knowledge. While this mysterious "agent" performs its office intelligently, the animal is a mere machine, according to the theory of Descartes. In other words, he is made a dwelling for a principle of intelligence; but this principle being superior to, and in some way independent of, the mute, holds no other relation to him than that of master and guide. Can anything be found in the whole range of human speculation more feeble than this expedient of human reason to explain a class of phenomena as simple as the simplest in the natural world?

[. . .]

The beaver, in a comparative estimate, is a low animal in his structural organization, as has been shown. He lives upon the coarsest food, is slow of motion upon land, of low respiration, monotrematous, and aquatic. His vision is short in range, and his brain is without those convolutions which are regarded as indications of mental power. In the great catalogue of animals, which is constructed upon the basis of anatomical structure, he rises no higher than the rat, the porcupine, or the squirrel. There is no reason for supposing that he is more intelligent than any other rodent of a corresponding grade. And yet by his sagacity, his industry, and his artificial erections, he has raised himself to a very respectable position, in human estimation, for intelligence and architectural capacity. It is because he needs these erections to promote his comfort and safety that man is able to follow the evidences of his skill and intelligence, and to become satisfied of their extraordinary character. If then an animal, with such an inferior organization, manifests so large an amount of mental capacity, of how much more must those be capable whose organization is found to be so much superior!

There is no doubt that the highest forms of intelligence among the mutes are to be found in the carnivorous animals. As an order they live pre-eminently by their wits; and they are unquestionably endowed with mental capacities, of higher relative power, to enable them to maintain their existence. The propagation and perpetuation of their species to the present time, testifies to the continuous triumph of their superior intelligence over the feebler capacities of the non-carnivorous mutes upon whom they subsist. They are able to endure hunger and fatigue, to wait and watch for prey, and to invent and practice many artifices for the capture of the latter. Many of them have great physical strength, a large brain, powerful respiration, and remarkable fleetness of foot. Their personal appearance commands both respect and admiration. Who ever looked into the clear round eye of a lion, without being impressed with the thought that there was a quick intelligence and a powerful will behind it, which, in the open plain or in the thicket, it would be hard to deceive and difficult to overmatch!

[. . .]

The several acts of the mutes here cited, as illustrations of the exercise of reason, can be fully explained as manifestations of a thinking principle. When the possession by them of such a principle is recognized, all difficulties vanish; and their conduct appears in an intelligible light. It also follows that their intelligence must necessarily be free to act within the range of its powers. In this discussion the relative strength of their mental capacities is left out of view, as immaterial. Compared with those of the human intellect they are feeble and slight, but within their several spheres of life and action they are ample for the promotion of their individual happiness.

V. *Imagination.* Whether the animal mind exhibits the quality of imagination it may be difficult to substantiate. Although it is one of the highest qualities of the mental principle, yet it is manifested in many simple forms. The playfulness of childhood, which is also commonly exhibited by the young of animals, is superinduced, seemingly, by the pictures or images formed in the mind by the fancy or imagination. This faculty, Kames[5] observes, "is the great instrument of recreation." If an attempt is made to explain the songs of birds, it will be necessary to resort to imagination, since the art itself is imaginative. Animals are known to dream from physical indications during sleep, and dreams are the works of memory and imagination. Too little is known of the lives of animals to show whether they possess this quality in any sensible degree.

VI. *The Will.* A doubt has been entertained whether the mutes possess a will, like the will of man, because responsibility must follow its exercise. Their own lives, at least, are intrusted to their keeping, the preservation of which is the highest form of responsibility. With a free volition, they rise up or lie down; they go or come; they play or quarrel, they bark, or mew, or sing; and they lie in wait for prey, or seek it by long excursions. These several acts are performed under the influence of motives, and were preceded by an exercise of the will. Unless the mute has a free choice between alternative courses, one of which may lead to danger and the other to safety, his conduct would be unintelligent. He might lose his life at any moment. The will is that mental power that sets the body in motion to execute a resolution previously reached by a process of reasoning. It is the power which adopts and executes the conclusions of the judgment. Unless a difference can be discovered in the quality of the will, as displayed by the mutes and by mankind, there is no means of distinguishing one from the other, except in the degree of its strength and persistency. A will, also, presupposes the existence of a mental principle, of which alone it can be predicated.

VII. *Appetites and Passions.* The mutes have the appetites and passions in common with mankind. No difficulty has ever been found in conceding a community of characteristics in these, the inferior, manifestations of the mental principle. While they differ in the degree of their strength, some of them are undoubtedly wanting among the lowest grades of the vertebrate animals. As a portion of them excel mankind in the acuteness of the senses, by means of which the feebleness of their mental powers is supplemented, so in some of the appetites and passions they may possess a delicacy of sensibility of which the human species are incapable. In their affections for their young, and for their mates (among such as pair), the highest evidence of their sensibility is found. They also display courage, fidelity, and gratitude, and to these, perhaps, in some rare instances, benevolence may be added. For the possession of these qualities, which are undistinguishable from the corresponding qualities manifested by the human mind, and for the beautiful illustrations of maternal affection which they display, they are entitled to our regard.

[. . .]

From the foregoing but most incomplete and imperfect consideration of some of the branches of the subject of Animal Psychology, it would be venturesome to urge any other than the more simple conclusions. Two or three only will be suggested.

In the first place, the term "instinct," to explain the intelligent acts of animals, should be abandoned. This term was an invention of the metaphysicians to assert and maintain a fundamental distinction between the mental principle of the human species and that of the inferior animals. With its multiform definitions, and with the repeated enlargements of its signification, it is wholly incapable of explaining the phenomena of animal intelligence. As a vain attempt to embody a system of philosophy in a definition, it has proved a failure, as might have been expected. When carried to its legitimate results, it endows every animal with a supernatural principle, and makes each of his intelligent acts little short of a miracle. As a term or invention it is *functus officio.*[6] With its disuse the subject of Animal Psychology is freed from all extraneous embarrassments, and the mental phenomena manifested by the mutes can be investigated and explained on philosophical principles.

In the second place, we are led to recognize in the mutes the possession of a free intelligence. In other words, that they are endowed with a mental principle which performs for them the same office that the human mind does for man; that this principle is free to act in view of motives and premises; and that it is ample in measure to enable each animal, within his sphere of action, to preserve his life and govern his conduct. This conclusion seems necessarily to follow from their possession of the organs of sense, from their manifestation of the appetites and passions, and from their ability to perceive, to remember, to reason, and to will.

And in the third and last place, as we are unable, in similar specific acts, to find any difference in kind between the manifestations of perception, appetite and passion, memory, reason and will on the part of a mute, and the corresponding manifestations on the part of a man, we are led to the conclusion that the difference is one of degree, and not of kind; and therefore that the principle from which they emanate is the same in kind, but bestowed in different measure, to adapt each species to its particular mode of life.

[. . .]

It cannot be said that the views, herein presented, tend to lower the personal dignity of man; but, on the contrary, they rather serve to distinguish his position. His great superiority is abundantly assured by the bestowment of the highest structural organization, of the fullest mental endowments, and by the possession of articulate speech. The distance which separates him from the highest of the mutes is sufficiently immeasurable to relieve his pride from all sense of humiliation from the consciousness of sharing the principle of intelligence with the latter. Sidney Smith[7] has touched this point with his satirical pen in the following language: "I confess I feel myself so much at ease about the superiority of mankind – I have such a marked and decided contempt for the understanding of every baboon I have ever seen – I feel so sure that the blue ape without a tail will never rival us in poetry, painting, and music, that I see no reason whatever that justice may not be done to the few fragments of soul and tatters of understanding which they may really possess." The mental principle here derided, while its possession is admitted, has, nevertheless, the inherent dignity of which a thinking principle cannot be divested. By his pre-eminent endowments, man stands at the head of the animal kingdom, the great exemplar of this principle, and separated by a wide interval from its other possessors. The separation is as marked and real as could be desired. But it is doubtful whether he possesses the sum of the powers of the principle called mind. It is precisely here, as it seems to the writer, that God has revealed a feature in the plan of creation not less wonderful than the original conception of a mental principle. Having called into existence this marvelous principle, and created a series of organic forms, He apportioned it among them all in such measure as to adapt each individual being to the sphere of life in which he was designed to move. The widest possible range for the exercise and development of mind was thus provided. A full comprehension of its powers and capacities must therefore be sought in its varied manifestations by the several species. It is not probable that the whole of its powers are possessed by any species: but rather that in their totality they are to be found among the members of the animal kingdom as a whole. A true system of mental philosophy, therefore, cannot be developed until all the manifestations of this principle are comprehended.

The hiatus between man and the nearest species below him in the scale of intelligence is so wide as to disturb the symmetrical gradation of the several orders of animals. We can neither conjecture that some intermediate order has fallen out of existence, nor assume the permanent degradation of any existing species; but, on the contrary, it seems to have been a part of the original plan of creation that man should stand without a compeer or contestant, the indisputable head of the series of organic forms, and the recipient, in the largest measure, of the gift of the mental principle. Some explanation of his excessive superiority may be found in the progress he has made since his emergence from his primitive condition. For ages, the bounds of which are unknown, mankind were immersed in a barbarism the depths of which can be but feebly conceived. They were without arts, without agriculture, without flocks or herds, depending chiefly upon fish, and the spontaneous fruits of the earth for subsistence. There are glimpses afforded to us, here and there, of a state of society in which the family relations were unknown, and in which violence and passion reigned supreme. The contrast between such a condition of mankind, and that of the present time, is so great that it is difficult to recognize in these primitive barbarians our lineal progenitors. Out of that condition man has struggled through a long and painful experience until he has been finally rewarded with the amenities of civilization. Language has been the great instrument of this progress, the power of which was increased many fold when it clothed itself in written characters. He was thus enabled to perpetuate the results of individual experience, and transmit them through the ages. Each discovery thus became a foundation on which to mount up to new discoveries. With the knowledge he has gained, and the elevation he has experienced, it is now difficult to realize the low condition from which his line of advancement commenced. Portions of the human family are still found in the darkness of ignorance, and in the feebleness of mental imbecility; and yet, although the distance of their intellectual separation is very great, it is much less than that between the latter and the most intelligent of the inferior animals. The difference expresses the superiority of his structural organization and of his mental endowments.

On the other hand, can it be truly affirmed that the inferior animals have been stationary in their knowledge from the commencement of their existence? This conclusion should not be over-hastily assumed. Within the period of human observation, their progress has seemed to be inconsiderable – but yet not absolutely nothing. For example, dogs under training have developed special capacities, such as the pointer and the setter, and have transmitted them to their offspring. This shows not only progress, but that of so marked a character as to work a transformation in the characteristics of the animal. Many animals, as the elephant, the horse, the bear, and even the hog – the type of stupidity – have been taught a variety of performances, under the stimulus of rewards, of which they were previously ignorant. These examples, however, are less important than the knowledge acquired by undomesticated animals, and transmitted, as a part of their experience and knowledge, in the species in which they were acquired. Of this kind are the several varieties of the beaver lodge and dam, and the development and perpetuation of the idea of a beaver canal. When careful and patient investigation has been made of these several subjects, the results will materially modify, in all probability, our present impressions.

Finally, is it to be the prerogative of man to uproot and destroy not only the masses of the animal kingdom numerically, but also the great body of the species? If the human family maintains its present hostile attitude toward the mutes, and increases in numbers and in civilization at the present ratio, for several centuries to come, it is plain to be seen that many species of animals must be extirpated from the earth. An arrest of the progress of the human race can alone prevent the dismemberment and destruction of a large portion of the animal kingdom. Domestication or extermination is the alternative already offered not alone to species, but to families and orders of animals. It may be that this result was never intended in the councils of Providence. It is not unlikely that God has adjusted a balance among the several orders of animals which cannot be overthrown except at the peril of the aggressor; and that in some mysterious way this balance is destined to be preserved. The present attitude of man toward the mutes is not such, in all respects, as befits his superior wisdom. We deny them all rights, and ravage their ranks with wanton and unmerciful cruelty. The annual sacrifice of animal life to maintain human life is frightful, if considered only with reference to its excess beyond our reasonable wants. When the Creator made man omnivorous, He designed his use of animal food. It is not sentimentalism but rather sense, to say that he should exercise the right with reason and forbearance. When we claim that the bear was made for man's food, we forget that man was just as much made to be food for the bear; and that our right to eat the bear rests upon no higher sanction, than his coequal right to feast upon our flesh if he overcomes in battle. Man's dominion over the mutes is in virtue of his superior endowments; but it is equally clear that the great Author of existence designed the happiness of the smallest and least endowed of all His creatures as completely and as absolutely as He did the happiness of man. If we recognize the fact that the mutes possess a thinking, and reasoning, and perhaps an immortal principle, our relations to them will appear to us in a different, and in a better light.

Editorial Endnotes

1 *Abercrombie*: John Abercrombie (1780–1844) was a leading Scottish physician who began his career studying the pathologies of the brain and nervous system. He ran a highly successful medical practice in Edinburgh, but later in his life shifted his attention to philosophy.

2 *Sir William Hamilton*: Sir William Hamilton (1788–1856), was a Scottish philosopher known for his contributions to the field of metaphysics, epistemology, and logic. He taught at the University of Edinburgh, where he first held a professorship of civil history and later, and more influentially, a professorship of logic and metaphysics.

3 *"that a plurality of principles . . . can possibly be explained by one"*: the principle known as Occam's razor – attributed to medieval Scholastic philosopher William of Ockham – which holds that in scientific explanations "plurality should not be posited without necessity."

4 *Dr. Reid*: Thomas Reid (1710–1796), a Scottish philosopher known for his common-sense philosophy and theory of perception. He taught at King's College Aberdeen and at the University of Glasgow, where he succeeded economist and philosopher Adam Smith as Glasgow Professor of Moral Philosophy.

5 *Kames*: Henry Home, Lord Kames (1695–1782), an important figure of the Scottish Enlightenment whose writings spanned the fields of social theory and legal and moral philosophy.

6 functus officio: without authority or power, indicating that a person or body has fulfilled its purpose and therefore no longer has any legal authority or power in the matter.

7 *Sidney Smith*: Sydney Smith (1771–1845) was an English writer, clergyman, parliamentary reform advocate, and one of the founders, in 1802, of the influential literary and political magazine *Edinburgh Review*.

Editorial Headnote

5. Arthur E. Brown, Primate Minds and Morals in the Philadelphia Zoo (1878–1883)

5.1 Arthur Erwin Brown. "The Serpent and the Ape," *The American Naturalist* 12, no. 4 (April 1878): 225–28.
5.2 Arthur Erwin Brown. "Grief in the Chimpanzee," *The American Naturalist* 13, no. 3 (March 1879): 173–75.
5.3 Arthur Erwin Brown. "The Kindred of Man," *The American Naturalist* 17, no. 2 (February 1883): 119–30.

Arthur Erwin Brown (1850–1910) was a Pennsylvania herpetologist who began working for the Zoological Society of Philadelphia in 1876, during the summer and fall of which the city became home to millions of visitors to the Centennial Exposition. A fair share of those visiting the world's fair were also drawn to Philadelphia's recently opened zoological garden, for which the then only 26-year-old Brown had just began working as director, a position he would hold until his death 34 years later. In 1897, Brown was elected Secretary of the Philadelphia Zoological Society and, at the time of his death, also served as vice president of the Philadelphia Academy of Natural Sciences and on the board of the Wistar Institute of Anatomy and Biology, located in the same city.[1]

While Brown was mostly known in the scientific community as an authority on snakes, his role as "custodian of the garden" also gave him ample opportunity to study the behavior of other creatures than his customary reptilians and amphibians. Brown published numerous articles in the *Proceedings of the Academy of Natural Sciences of Philadelphia*, *Science*, and the *American Naturalist*, a leading American journal of biology founded in 1866 by four former students of Harvard zoologist Louis Agassiz who had fallen out with their influential mentor.[2] It is in the *Naturalist* that Brown published some of his personal observations on the behavior of primates in Philadelphia Zoo combined with general thoughts on primate mental capacities and moral "character." The focus of many of these observations was a pair of chimpanzees who, in a humorous nod to post-Darwinian debates around human origins, had been christened Adam and Eve and now lived in the zoo's Monkey House. Brown's final article for the *Naturalist*, "The Kindred of Man," includes his most comprehensive deliberations on the subject.

Notes

1 Roger Conant, "Arthur Erwin Brown: 'Custodian of the Garden' and Naturalist of Note," *America's First Zoo* 9, no. 4 (December 1959): n.p.; Zoological Society of

Philadelphia, *The Thirty-Ninth Annual Report of the Board of Directors of the Zoological Society of Philadelphia* (Philadelphia: Allen, Lane and Scott, 1911), 14.

2 Edwin G. Conklin, “The Early History of the American Naturalist,” *The American Naturalist* 78, no. 774 (1944): 29–37; L. C. Dunn, “The *American Naturalist* in American Biology,” *The American Naturalist* 100, no. 915 (October 1966): 481–92.

5

PRIMATE MINDS AND MORALS IN THE PHILADELPHIA ZOO (1878–1883)

Arthur E. Brown

Sources: Arthur E. Brown, [Primate Minds and Morals in the Philadelphia Zoo], *The American Naturalist* (1878–1883)

5.1 Arthur Erwin Brown. "The Serpent and the Ape," *The American Naturalist* 12, no. 4 (April 1878): 225–28.

5.2 Arthur Erwin Brown. "Grief in the Chimpanzee," *The American Naturalist* 13, no. 3 (March 1879): 173–75.

5.3 Arthur Erwin Brown. "The Kindred of Man," *The American Naturalist* 17, no. 2 (February 1883): 119–30.

5.1 THE SERPENT AND THE APE.

BY ARTHUR E. BROWN.

WITH the purpose of observing the manner in which the feelings of curiosity and astonishment are manifested in the monkey, Mr. Darwin once introduced a snake into a cage containing a number of those animals, and the results of his experiment he refers to in "The Descent of Man," and also in "The Expression of the Emotions in Man and Animals," as illustrative of the extent to which those qualities are developed in that branch of the animal kingdom.

Reading his statement, the writer conceived the idea that the results obtained were capable of a deeper application than was then given them, and he proceeded to try the experiment for himself. The Monkey House at the Philadelphia Zoological Garden afforded the opportunity, so a dead snake was coiled up in a newspaper, the corners of which were twisted together in such a manner that they would readily come undone, and the package was then set on the floor of a cage containing forty or fifty monkeys of a great variety of species. It was instantly spied by a female *Cynocephalus,*[1] who was the principal leader in all the pranks with which the monkeys constantly amused themselves; she seized the paper by one corner, and set off across the cage, dragging it behind her, evidently intending to have a good time with it.

Before she had gone more than a few feet, the paper became unfolded, and the snake slipped partly out. She instantly dropped the paper and sidled off in a very comical manner with her head over her shoulders, keeping an eye behind her,

DOI: 10.4324/9781003032717-6

much as Lot's wife[2] must have looked back on the fascinating terrors of the cities of the plain. No sooner did the rest of the monkeys perceive the dreadful object in their midst, than they approached, step by step, and formed in a circle of six or eight feet diameter, having for its centre the snake quietly coiled up on the floor. None dared, however, to touch it or to go beyond the established line of safety, with the exception of one large Macaque, the acknowledged leader of the cage, who cautiously approached and made an occasional snatch at the paper, apparently to see if the enemy was really as devoid of life as it appeared to her; all the others, meanwhile looking on in breathless attention.

At this point, a string which had previously been attached to the tail of the snake was gently pulled. The serpent lengthened slightly, and the monkeys fled up the sides of the cage, chattering and screaming like magpies; when they got to a safe distance they halted for observation, and after some moments, seeing no further sign of danger, they gradually returned, one by one, to their former position – the large ones in the front rank, and the smaller ones, crowded out by superior strength, forming behind and looking over their shoulders.

This was continued for some hours without the slightest change in the disposition of the monkeys – all of their actions showing a most intolerable fear of the snake, mingled with an attraction or curiosity which would not allow them to remain away from it. This was so universal that not one of the monkeys in the cage was entirely free from it.

The snake was finally taken out, and several other animals belonging to the same class were put in its place, but with very different results. Of a tortoise, for instance, and a small dead alligator, they were at first rather shy, but they at length began to touch them, and in ten minutes they were playing with them, and passing them from one to another with the greatest curiosity.

The same snake was then shown, in turn, to animals belonging to a number of other orders: Carnivores, Rodents, Ungulates, Edentates[3] and Marsupials, but none of them paid it any special attention with the exception of a Peccary (*Dicotyles labiatus*), which, finding it to be dead, seemed disposed to make a meal of it.

Turning from the monkeys and watching, instead, the visitors to the Reptile House, it is evident that the instinctive fear and horror of the snake which is so common as to be almost universal with man, is closely allied to that which has been seen to exist among monkeys. Women readily develop this, as their emotions are more quickly responded to by gestures, than is usually the case in the other sex, and I was specially fortunate, a short time after the occurrence detailed above, in having an opportunity of observing the effect produced by the collection of snakes, upon a lady who was deaf and dumb – by the fact of her disabilities she was shut out, to a very great extent, from the influence – repressing, so far as the expression of the emotions is concerned – of free association with others, and the nature of her feelings was thereby rendered more evident I was not at all surprised to trace in her, actions and gestures which resembled closely those which I had observed on the part of the monkeys; they evidenced the same fear, the same attraction and the same repulsion, and after watching for a long time, with an expression of the most intense disgust, the cage of Boas, she was at last led away by her friends, protesting that she wanted to stay.

Now if it be asked why this instinctive feeling should be developed in the *Primates* alone – it is probable that as the early dawnings of intelligence in the common ancestor of man and monkeys began to surpass the power of receiving impressions which existed in other animals, he would be most liable to conceive great dread of that enemy which inflicted upon him wounds of a very different sort from those which he received from his own kind or from animals which approached more or less to his own form, and which also produced effects so subtle in their character and operation, that they would be apt to leave lasting impressions on those animals which were frequently subjected to witnessing them. It should be remembered, also, that the home of the monkey and the spot where, in all probability, the earlier *Primates* first saw the light, is in those regions of the earth which are most infested by numerous and venomous serpents.

These facts will at once suggest to all who put their faith in the theory of gradual development, that the fear of the serpent became instinctive in some far distant progenitor of man, by reason of his long exposure to danger and death in a horrible form, from its bite, and that it has been handed down through the diverging lines of descent which find their expression to-day in *Homo* and *Pithecus.*[4] How strongly marked it is in the latter, the experiment detailed above, corresponding in each of its results with that of Mr. Darwin, bears testimony; and for the evidence of its influence on the mind of the former, turn to the story of the serpent in Paradise; to the signs and symbols of many ancient mythologies, and to the feeling which few men can deny to themselves when they are brought into association with even the most beautiful and harmless member of the order *Ophidia.*

5.2 GRIEF IN THE CHIMPANZEE.

BY ARTHUR E. BROWN.

SOME months ago I called attention in the "Notes" of the NATURALIST to several evidences of a high degree of mental power on the part of the chimpanzee. One of the pair which, at that time, was in the Philadelphia Zoological Garden, has since died, and the behavior of the surviving one on that occasion appears to me to bear somewhat on the acquired nature of the physical means by which our strongly excited emotions find relief, as well as on the origin of those emotions themselves.

Among the lower animals, with the exception of some domesticated varieties, any striking display of grief at the death or separation from an animal to the companionship of which they had been accustomed, has rarely been observed, and although a few statements of such occurences have been made by different authorities, it is probable that the feeling of individual association, or friendship – if the term maybe so used – partakes too much of an abstract nature to be sufficiently developed in them to retain much of a place in memory when the immediate association be once past. This would seem to be the case even in one of the strongest of animal attachments – the maternal instinct – in which the direct presence of the offspring, acting as a stimulus, calls forth the emotion of the mother, which, strongly rooted as it appears to be, contains much of a reflex nature and ceases on the disappearance of its cause. And here let it be said, that although the instinct of maternity and the sentiment of friendship perhaps differ widely in their origin, yet in their manifestations they are so nearly alike that the reverse feelings excited by any violence done to them, need not and probably do not differ much in kind.

With the chimpanzee, the evidences of a certain degree of genuine grief were well marked. The two animals had lived together for many months, and were much attached to each other; they were seldom apart and generally had their arms about each other's neck; they never quarreled, even over a pretended display of partiality by their keeper in feeding them, and if occasion required one to be handled with any degree of force, the other was always prepared to do battle in its behalf on the first cry of fright. After the death of the female, which took place early in the morning, the remaining one made many attempts to rouse her, and when he found this to be impossible his rage and grief were painful to witness. Tearing the hair, or rather snatching at the short hair on his head, was always one of his common expressions of extreme anger, and was now largely indulged in, but the ordinary yell of rage which he set up at first, finally changed to a cry which the keeper of the animals assures me he had never heard before, and which would be most nearly represented by *hah–ah–ah–ah–ah,* uttered somewhat under the breath, and with a plaintive sound like a moan. With this he made repeated efforts to arouse her, lifting up her head and hands, pushing her violently and rolling her over. After her body was removed from the cage – a proceeding which he violently opposed – he became more quiet, and remained so as long as his keeper was with him, but catching sight of the body once when the door was opened and again when it was carried past the front of the cage, he became violent, and cried for the rest of the day. The day following, he sat still most of the time and moaned continuously – this gradually passed away, however, and from that time he has only manifested a sense of a change in his surroundings by a more devoted attachment to his keeper, and a longer fit of anger when he leaves him. On these occasions it is curious to observe that the plaintive cry first heard when the female died, is frequently, though not always made use of, and when present, is heard towards the close of the fit of anger. It may well be that this sound having been specialized as a note of grief, and in this case never having previously been called into use by the occurrence of its proper emotion, now finds expression on the return of even the lesser degree of the same feeling given rise to by the absence of his keeper, and follows the first outbreak of rage in the same manner as the sobbing of a child is the natural sequence of a passionate fit of crying. It may be noted too, that as his attachment to his keeper is evidently stronger than when there was another to divide with him the attention which they received, the grief now caused by the man's absence would naturally be much stronger and a more exact representation of the gestures of grief would be made.

Notwithstanding the intensity of his sorrow at first, it seems sufficiently evident that now a vivid recollection of the nature of the past association is not present To test this a mirror was placed before him, with the expectation that on seeing a figure so exactly like his lost mate, some of the customary signs of recognition would take place, but even by caressing and pretending to feed the figure in the glass, not a trace of the expected feeling could be excited. In fact, the only visible indication of a change of circumstances is that while the two of them were accustomed to sleep at night in each other's arms on a blanket on the floor, which they moved from place to place to suit their convenience, since the death of the one, the other has invariably slept on a cross-beam at the top of the cage, returning to inherited habit and showing, probably, that the apprehension of unseen dangers has been heightened by his sense of loneliness.

On looking over the field of animal emotion it seems evident that any high degree of permanence in grief of this nature belongs only to man; slight indications of its persistence in memory are visible in some of the higher animals and domesticated races, but in most of them the feeling appears to be excited only by the failure of the inanimate body, while present to the sight, to perform the accustomed actions.

The foundation of the sentiment of grief is probably in a perception of loss sustained in being deprived of services which had been of use. An unrestrained indulgence in an emotion so powerful as this has become in its higher forms, would undoubtedly prevent due attention to the bodily necessities of the animal subjected to it; in man, its prostrating effects are mainly counteracted by an intelligent recognition of the desirability of repairing the injury suffered, and in him, therefore, the feeling may exist without serious detriment to his welfare, but among the lower animals it would seem probable that any tendency to its development would be checked by its own destructive effects – the feeling, for instance, would most frequently occur on the death of a mate – a deep and lasting grief would then tend to prevent a new association of like nature and would thus impede the performance of the first function of an animal in its relation to its kind – that of reproduction.

5.3 THE KINDRED OF MAN.

BY ARTHUR ERWIN BROWN.

MR. A. R. Wallace[5] once called attention to the similarity in color existing between the orang and chimpanzee and the human natives of their respective countries. It would indeed seem as if but half the truth had been told, and that the comparison might be carried also into the region of mind; the quick, vivacious chimpanzee partaking of the mercurial disposition of negro races, while the apathetic, slow orang would pass for a disciple of the sullen fatalism of the Malay.[6] Such, at least, was the impression left by careful observation of several specimens of each species which have been exhibited in the Philadelphia Zoölogical Garden.

A curious study are the moral qualities of the chimpanzee – for he has morals – not altogether such as would serve for the ordering of a human community, but very well adapted, seemingly, for his own needs. Watching them closely, in all their moods, all their passions, it was impossible to avoid the feeling that here was man in his primitive stage of moral development – "nature's ground plan" only – self-love predominant, the brute mainly, with but an occasional flash of the possibilities which the hand of nature was yet to shape.

"Adam" and "Eve" were both young, probably not more than three or four years old, and not half grown, as the chimpanzee is believed to require some twelve or fifteen years for the completion of that stage of existence. They were about the same size – perhaps they were twins – they had no family Bible to settle the question, but the extraordinary likeness between them was strongly in favor of the supposition; indeed, if Adam had not been ornamented with a black smudge across the nose, they could hardly have been told apart; but twins or not, they loved each other with a most devoted affection, or, at least, so it seemed, but subsequent events cast a doubt on the real depth of their feelings.

Being young they were eminently social, for it may be said that as a rule, among wild animals, moroseness and ferocity come only with age. When they were first coaxed out of their traveling cage they were visibly embarrassed, and retired into the nearest corner, locked tightly in each other's arms, which, as we afterwards learned, was a universal refuge in time of doubt, but it was not long before they began to feel at home, and thenceforward were always ready to make friends with anybody who made his approaches in due form. As has been said, they were very fond of each other, and it was on rare occasions only that they were not clasped in a fond embrace, and not once during their life in the garden was anything but the most perfect accord manifested between them. No pretence of partiality in feeding, no petting of one to the exclusion of the other, could excite a trace of jealousy; the slighted one would simply retire to a corner and sulk, but their mutual relations were undisturbed; resentment was all towards the giver, not to the one who received. Each was at all times ready to stand by the other; probably the keeper has not yet forgotten the ferocious assult Eve once made on him from the rear, while he was engaged in pouring a dose of medicine down the throat of her companion.

Their anger was something ludicrous; the male especially was liable to paroxysms of rage, during which he would tear his hair with both hands, hurl himself down on the floor with a perfect tempest of yells and roars, but in a moment it was all over, and he was ready to make peace and accept any small attention by way of *amende.* But his masculinity asserted itself more particularly when danger seemed to threaten – then he was grand; advancing inch by inch, brandishing his arms, stopping after each step, with a stern frown, to emit a terrifying roar, he seemed an impersonation of resolution and defiance – a very Ajax[7] – but after all, he was only a Thersites,[8] a more arrant little coward than he, at heart, had never lived, and if his appearance did not have the desired effect, if the intruder stood his ground, the dignified approach gradually became slower, the resolution ebbed away, and the inevitable end was a final stop, a hasty turn and an ignominious flight into the corner – generally, it is painful to say, behind Eve. We could only blush, we dared not blame him; one nearer to us far than he, his namesake, under circumstances which brings the action home to each of us, had done the same.

Many experiments were made to test the mental capacity of these animals, with quite fruitful results; the primary mental operations, and even some which involved a greater or less combination of ideas, were performed by them with facility; indeed, it may be doubted if the undirected efforts of a human child of the same age, ignorant of language, could produce results of a much higher grade.

A mirror being placed in the cage, the male, after cautiously investigating the figure reflected, turned it over, and finding nothing but bare boards, he placed it face downwards on the floor and executed a sort of war dance on the back. Having repeated this a number of times, the glass was held firmly before him; he then gave it more attention, at first attempting to drive away the figure he saw; at last a resemblance seemed to strike him, and after performing a variety of antics, seemingly for the purpose of comparison, it was quite evident that he became aware of his own identity – and in this, perhaps, afforded a trace of that self consciousness which conservative philosophy allows only to the lordly intellect of man. In this case it is probable that he had become accustomed to see a faint image of himself reflected from the glass front of his cage.

Perhaps, though, the most striking evidence of their power of reasoning was given when a dead snake was taken into the room and shown to them. As is well known, monkeys have an intolerable dread of snakes, and these were terrified beyond measure. They fled at once to the highest point of the cage, uttering their expressive cry of fear, and there they remained for hours, refusing to come down even at sight of their accustomed dish of food, and when at last they did so, it was with the greatest caution; a slight movement in the straw covering the floor, was enough to cause a panic, and it was some hours before they fully recovered, equilibrium. Finally, when both were sitting quietly near the glass front, the snake was suddenly shown to them on the outside, but there, the object which had caused such terror in the same room, was powerless; the glass which they themselves could not pass, was a barrier as well to their enemy, and they simply sat still, pointing at it with their fingers and uttering the *hoo-hoo* which expressed doubt, dislike and disgust. It was suspected that they had only become accustomed to the sight, and to complete the experiment the snake was again thrown in through the back door, when the terror of the two animals was as great as on the previous occasion.

It was quite clear that they possessed a limited means of vocal communication. Sounds, to the number of three or four, were uttered by one, which met with a different response from the other, either by voice or action, and in which it was quite possible for the human ear to detect a difference.

The imitative habits common to the whole tribe of monkeys were strongly marked in them, and made it relatively easy to teach them to use a spoon or drink out of a cup, and to perform various small tricks. In cleaning the glass in front of the cage with paper, the keeper generally threw two pieces on the floor beside him, when each chimpanzee would take one of them and set to work polishing the glass in like manner, deriving, apparently, great satisfaction from the performance. The delicacy of their taste became developed to an extraordinary degree by the varied diet afforded them; both were fond of the taste of sherry, which was always put in their tapioca and cornstarch, but when brandy, whisky and rum were substituted, they stuck out their under lip in disgust and refused to eat it.

For music they had no ear whatever; the notes of an accordeon and violin produced in them only distrust of the instrument, and when these were put into their hands, their insatiable curiosity prompted only efforts to find out what was inside.

For many months Adam and Eve were the pets of the "Zoo;" few days there were when they were not surrounded by a crowd of interested spectators, some of whom, to a critical eye, were fully as amusing as the animals they came to see.

It is learned from African travelers that the native tribes inhabiting the range of the gorilla and chimpanzee believe them to be human beings who have degenerated from their original state, and that out of pure laziness they refuse to speak, in the fear that if their possession of the faculty should become known, they will be set to work in the fields; indeed, the native name of the chimpanzee, *Enge-e-co,* means "hold your tongue," and evidently originated in this belief. It frequently seemed as if similar ideas prevailed among a certain part of the visitors, and that class especially whose acquaintance with the forms of orthography had not reached a familiar stage, seemed to find in the scientific name of the animal, *Anthropopithecus niger,* indications, of a relationship to the humble man and brother whose ancestors sprang from the same soil.

But at last, in spite of tender care and attention, Eve became sick – poor little thing, how she did suffer. Of course she ought to have been a good and grateful patient and have known that everything done for her was for her ultimate benefit – they always do in the animal literature of the day – but she had read little, and so was hardly to blame in following out the instincts of her nature. She might have been expected to look appealingly into the *eyes* that bent over her, but she did not; she ought to have pressed affectionately the hand that cut the hair from off the region of her little stomach and gently applied a mustard plaster to the affected part, but instead, she bit it savagely; and to crown all, she was so little sensible of the soothing influence of that mustard plaster that it took the united efforts of three men to keep it in place until its work was done.

Alas for all the works of fancy! a long experience of sick and suffering animals compels the conclusion that one of the things which is beyond the grasp of mind to be found among the lower animals, is surgery.

And so Eve passed out from the familiar places of the "Zoo." Her funeral urn stands ranged on a shelf in that universal mausoleum of nature, the Academy of Natural Sciences, and her "In Memoriam," by Professor Chapman,[9] was published in the Proceedings of that venerable institution.

Adam was left alone to mourn, but to his shame be it said that although he was inconsolable at first, so long as the dead body of his late companion was in sight, he soon got over it, and in forty-eight hours not a trace of her seemed to exist in memory, excepting that to the day of his death, some months later, he was afraid to sleep alone on the floor, where the two had always slept together, and with the shades of night he followed out his ancestral habit, climbed as high as he could get towards the roof, and there composed himself to peaceful slumbers.

For some time the garden was without any specimen of the higher apes, until in the autumn of 1879 a young orang-utan was safely received.

There is something about the orang that irresistibly suggests a spider – one of those red, hairy, long-legged spiders which one sees with an instinctive feeling of repulsion. At no age can the animal be called handsome, and the old males, covered with coarse, reddish-brown hair six or eight inches long, with a huge protruding jaw and a mass of hardened skin on each cheek, are about as unprepossessing as anything that nature has produced. "Topsey," however, as is sufficiently indicated by her baptismal name, belongs to the fairer sex; her age – probably for that reason, is unknown. When she arrived she was supposed to be about two or three years old, but as the lapse of time has made hardly any change in her personal appearance, save in the way of *embonpoint,*[10] it is probable that she was older, although she is certainly not half grown; if, indeed, as has been suggested, she may not be a dwarf – a sort of feminine Tom Thumb[11] among orangs; and in this, possibly is the explanation of the unusually good health which she has enjoyed through a lifetime much longer than is common to her species in captivity. The amount of nutrition required to simply maintain the existing condition of body, would of course be less than if the processes of growth were in full activity, and the assimilation of food, which is probably defective in most caged animals, would, as has been the case here, be sufficient to keep her in good condition.

Between the orang and chimpanzee there is a marked difference in moral qualities. The latter is full of life, vigor, vivacity; lively and child-like in disposition, enjoying life to the full, and taking interest in all that goes on about it. Quite the reverse with the orang – it is slow, sluggish and calculating; philosophically indifferent to everything but its immediate wants – voluptuary and stoic in one – life is only for the means of living, and life itself is hardly worth the pain of an exertion. It is exasperating – the apathy of the orang; for hours it will lie wrapped in a blanket close to the front of the cage, lazily following with its eyes the motions of any person who comes within its range of vision, or slowly blinking at a straggling fly upon the glass, moving – when it must move – only with the greatest deliberation. If left hanging by one hand to a rope or branch, there it will hang, perhaps for several minutes, before making up its mind to take hold with the other or let go altogether. Latterly the contrast in the disposition of these animals has been made very striking by the presence in one cage of specimens of each species. A second pair of chimpanzees, about the same size as the orang, were placed with her, and with their natural liveliness at once made overtures of acquaintance, which were as promptly repulsed, and during the first week she suffered so much fright and uneasiness from their perfectly good-natured attempts to induce her to join in their play, that it became necessary to partition off with wire screens a corner of the apartment, and there, hour after hour, while the two chimpanzees are climbing, swinging and tumbling about the cage, never at rest except to plan some new scheme of amusement, the orang lies flat on her back, fingers and toes closely interlocked in the air, enjoying a *dolce far niente,*[12] the relish of which she seems to intensify by quiet wonder at the reckless prodigality of force indulged in by her neighbors.

This stolidity is characteristic of the species in a wild state; there they live mostly in the tree tops, cautiously crawling from branch to branch, testing every limb before resting their weight upon it, moving only to satisfy the demands of hunger, and when that stimulus to action ceases, subsiding into a half-sitting position with the trunk or branch of a tree to hold up the back, head bowed on the breast, hands hanging down – not asleep – it can be nothing but laborious thought that produces such perfect bodily repose. Who can tell how deeply the meditative orang has penetrated into the mysteries of the cosmogony of which he is a part? how many systems of philosophy have dawned, after hours of reflection, into his weary brain? how deeply he has pondered on the origin and destiny of his race, and to how many metaphysical final causes has his speculative career traced its way?

The orang is really not so stupid as appearances would have it, and it is an interesting fact that the actions of the one in question once gave evidence – and the only evidence the writer has ever observed among the lower animals – of what seemed to be some understanding of death. Another orang had been procured as a mate, and arrived in bad health; it was exceedingly irritable, and though weak from disease, managed to appropriate the only blanket in the cage, and fought off the rightful proprietor whenever she approached. This, with other grievances, caused Topsey to regard the intruder with marked dislike and fear. She watched it from a distance all through the several days of illness, and the more attentively as the last moments drew near and pain and weakness were showing plainly their ravages, until finally, after a hard struggle, the little sufferer lay motionless and dead, then, for the first time, she drew near, looked at the body for a moment, pushed it with one hand, and then after putting her nose close down against its face, as if to listen for a breath or any sign of life, she began pulling from under it the coveted blanket which it was no longer able to defend, and in the most satisfied manner wrapped herself up and laid down in peace.

Much less opportunity has been afforded for critical observation of the remaining anthropoids – the gorilla and the gibbons – as few of either have been kept in captivity; but the former may fairly be considered as not presenting marked mental differences from the chimpanzee, and the latter seem in all respects to be below the level of the others.

[. . .]

Observation renders it quite probable that mental capacity in these animals has, to a considerable degree, maintained a relation to the complexity of detail in brain structure, although undoubtedly, from a mere comparison of human and anthropoid brains, a far greater degree of intellectual power than that which really exists, might be expected from the latter; it should be remembered, however, in favor of the ape, that the specimens from which our ideas of their intelligence have been derived, have for the most part, been very young, and it is possible that more mature age may bring with it a higher degree of mental faculty. On the whole, however, it is quite certain that the intelligence of the lower animals has been greatly overestimated. All experienced observers of their actions know how easy it is to place a motive and an understanding where none probably exist. It is difficult, except after long training, to withstand the influence of the subjective tendencies of the mind, which lead the observer to translate into the terms of his own intelligence, those actions which seemingly correspond to his own desires, and there are few works on this subject in which constant evidence is not given of its presence. In experimenting with the animals which form the subject of this paper, the difficulty was constantly met with, and a large proportion of the phenomena observed were set aside, reluctantly in many cases, because of the doubt.

[. . .]

It is held generally, in popular misconceptions of the doctrine of evolution, that man is a direct descendant of the higher apes, and the gorilla is commonly looked on as being his nearest progenitor. From the standpoint of science, however, no student of biology will maintain that the ancestry of man has yet been fully traced, but will limit himself to the conviction that at some period of the prehistoric world, the forces of nature, acting from without, on the plastic materials of life, have brought down from an unknown point of departure – perhaps among the lemurs – two diverging lines of development, one of which finds its present type in man, the other in the Catarrhine monkeys[13] and their highest form – the anthropoids.

Perhaps the future of science may unfold the details of development, but to do this it is probable that ages of geological upheaval will be required, to bring above the ocean continents long buried, in which the process took place and in which the records are contained.

Manlike as are the apes, there is a contrast which the resemblance serves, in great part, but to intensify – anatomy finds similarity throughout and takes note of little that is unlike, while function, based upon these structures, has become so specialized and elevated during progress from the lower to the higher, as to become almost difference, and man and ape are in fact as in time separated by a gulf so vast that the furthest reach of science can catch, as yet, but shadowy outlines of the other side.

Editorial Endnotes

1 Cynocephalus: Greek for "dog-headed," a term that came to be used in reference to the yellow baboon (*papio cynocephalus*).

2 *Lot's wife*: In the book of Genesis, Lot is Abraham's nephew who settles in Sodom. As Sodom faces destruction, Lot and his family are warned by angels to flee without looking back. Lot's wife disobeys this command, looks back at the city, and is turned into a pillar of salt.

3 *Edentates*: formerly an order ("Edentata") of toothless or nearly toothless mammals native to South and Central America that includes anteaters, sloths, and armadillos.
4 Pithecus: Latinised version of the Greek *píthēkos* ("ape").
5 *A. R. Wallace*: Alfred Russel Wallace (1823–1913), a British naturalist and geographer whose many scientific contributions include his development of a theory of evolution through natural selection that closely aligned with Darwin's and pushed the latter to publish his own ideas in *On the Origin of Species*.
6 *the Malay*: the idea of a "Malay race" as a collective term for various Indigenous peoples of Southeast Asia, especially in the Malay Archipelago, originated in the racial anthropology of German physician Johann Friedrich Blumenbach (1752–1840).
7 *Ajax*: a powerful Greek warrior and hero of the Trojan War who appears in Homer's epic poem *Iliad*.
8 *Thersites*: unlike Ajax, Thersites is described in the *Iliad* as a lowly, weak, and physically unattractive soldier who voices his opposition to the war effort before being confronted and humiliated by Odysseus.
9 *Professor Chapman*: Henry Cadwalader Chapman (1845–1909), a Pennsylvania physician and a member of the Philadelphia Academy of Natural Sciences and the Zoological Society of Philadelphia. His research on primate anatomy was facilitated by his connection to the Philadelphia Zoo, which provided him with research materials and specimens, including a gorilla and a chimpanzee for dissection.
10 embonpoint: plumpness or stoutness.
11 *Tom Thumb*: "General Tom Thumb," whose real name was Charles S. Stratton, was affected by a rare condition today known as pituitary dwarfism and one the first major attractions of notorious showman and circus pioneer P. T. Barnum.
12 dolce far niente: an Italian expression that translates to "sweet idleness" or "the sweetness of doing nothing."
13 *Catarrhine monkeys*: "down-nosed" monkeys, one of the two major groups of simians. Catarrhines are native to Africa and Asia (Old World monkeys) and have evolved independently from their New World counterparts, the platyrrhines.

Editorial Headnote

6. Francis Bowen, "The Human and the Brute Mind," *The Princeton Review* 56 (May 1880): 321–43.

Francis Bowen (1811–1890) was a Massachusetts philosopher who taught at Harvard University, where he held the Alford Professorship of Natural Religion, Moral Philosophy, and Civil Polity from 1853 to his retirement in 1889. Executing the will of its wealthy patron, John Alford, who had served on the Harvard Board of Overseers and specified that parts of his estate should be made available for "pious and charitable uses," the original 1789 endowment of the professorship specified that its holders – who had to officially declare their membership in the Protestant church for their inauguration – were to

> demonstrate the existence of a Deity or First Cause, to prove and illustrate his essential attributes, both natural and moral, to evince and explain his providence and government, together with the doctrine of a future state of rewards and punishments.[1]

Bowen himself certainly seems to have been committed to a defense of Christian theology in his publications on modern philosophy, metaphysics, and political economy. In a lengthy critical review of Darwin's *Origin of Species* for the April 1860 issue of the *North American Review*, Bowen claimed that Darwin's theory, "ingenious and captivating" as it was, lacked a foundation in empirical evidence, especially regarding its implication – or outright affirmation, in Bowen's reading of the book – "that the monkey is [man's] brother, and the horse his cousin, and the oyster his remote ancestor," and that the dignified human mind can be traced back in its development to the lowly sphere of animal instinct.[2] For Bowen, the answer to Darwin's speculations on origins and continuances lay in the "direct action of an intelligent creative cause."[3]

Bowen's anti-evolutionist stance, which made him a natural ally of fellow Harvardian and staunch anti-Darwinian Agassiz, still informs the following article for the *Princeton Review*,[4] published two decades after his review of *Origin* and nine years after the publication of Darwin's *Descent of Man*. In the article, Bowen again takes up the question of distinction and continuity between human and animal minds he had already been concerned with in his earlier engagement with the "heroic imaginations" of Darwin and other evolutionists.

Notes

1 Josiah Quincy, *The History of Harvard University*, vol. 2 (Boston: Crosby, Nichols, Lee & Co., 1860), 500, 504, 502. Biographical information on Bowen is taken from

R. Douglas Geivett, "Bowen, Francis," in *American National Biography*, ed. John A. Garraty and Mark C. Carnes, vol. 3 (New York: Oxford University Press, 1999), 276–77.

2 [Francis Bowen], review of *On the Origin of Species by Means of Natural Selection, or the Preservation of Favored Races in the Struggle for Life*, by Charles Darwin, *The North American Review* 90, no. 187 (April 1860): 474–506, here 482, 475.

3 Bowen, 504.

4 The *Princeton Review* was a theological journal that started in 1825 under the title *Biblical Repertory*. See Frank Luther Mott, *A History of American Magazines*, vol. 1: 1741–1850 (Cambridge: Harvard University Press, 1930), 529–35.

6

THE HUMAN AND THE BRUTE MIND (1880)

Francis Bowen

Source: Francis Bowen, "The Human and the Brute Mind," *The Princeton Review* 56 (May 1880): 321–29, 331–34, 336–38, 340–43.

THE question whether the human intellect differs from the brute mind in degree only, or also in essence and kind, is not new in the history of philosophy. It has been debated with earnestness in every age, at least since the revival of letters. In the sixteenth century, Montaigne, who is followed by Charron and Gassendi,[1] undertakes to prove that there is a wider interval of mental power between one man and another, than between man and the most sagacious brute. In his usual fleering and sceptical manner, he says, we must push man back into the crowd of animals from which, in the arrogance of his heart, he aspires to separate himself. In fact, so far as the animal acts from instinct, Montaigne declares it is the superior, since it then accomplishes at once, and without reflection or effort, tasks which man can perform only imperfectly and after repeated failures.

In the next century, Descartes went just as far into the opposite extreme, when he maintained that brutes are mere automata, destitute not only of intellect and feeling, but even of life. If the animal goes in pursuit of any object, he says, it is because an impression has been made upon its organs of sense, through which a spring is put in motion, that propels the beast in chase just as mechanically as the hands of a watch travel round its face after it has been wound up. Its inarticulate cries, which we wrongly interpret as signs of emotion, are, like the striking of a clock at determinate intervals, caused and regulated by its Internal machinery. This is the noted hypothesis of the animal-machines, which was discussed with so much spirit, both in prose and verse, especially in France, long after the death of its author. [. . .]

In our day, the disciples of Herbert Spencer,[2] and the evolutionists generally, take sides, of course, with Montaigne and against Descartes. As they hold that all modes of being and forms of life, from the lowest to the highest, are successively self-developed, through countless slight gradations, from the primitive atoms which are the formless elements of chaos, so they necessarily believe that all the faculties of the human mind exist also, tho in a rudimentary state, in the mental

DOI: 10.4324/9781003032717-7

constitution of the inferior animals, and may even be traced in imagination much farther back, to the mud or dust whence those brutes originated, In his "Descent of Man," accordingly, Mr. Darwin assures us that "the difference in mind between man and the higher animals, great as it is, is *certainly* one of degree and not of kind." In the affection of the dog for his master, for instance, he beholds the rudiments of the religious sentiment; in the social instincts, he finds the elements of morality; and in the inarticulate cries, "aided by gestures and movements of the muscles of the face," by which animals express their emotions, he detects the origin of language. And Mr. Huxley,[3] consistent fatalist as he is, contrives to unite the doctrine of Descartes with that of Montaigne, by maintaining that man also, like the dog, is an only seemingly animate automaton, and therefore does not essentially differ from the machine-brute; since the higher grade of evolution that he has reached sufficiently accounts for what appears to be the greater skill expended upon his construction.

Not much light is thrown upon the discussion of this subject by the marvellous stories, of which so many are current, of the signal forethought, feeling, and contrivance shown by particular animals on special occasions. Few of these anecdotes are so well authenticated as to deserve full credit; and they would not be reported but for their exceptional character. But only the habitual actions of the animal fully evince its real nature and capacities; feats which it may be trained to accomplish, and acts done under an unusual combination of circumstances, and seldom or never repeated, cannot be safely interpreted as proofs of intelligence. Most of them can readily be explained away, since we can hardly suppose that a brute is any wiser at one time than at another, or that one can be much distinguished for sagacity above his fellows. It is all a matter of interpretation and rather uncertain inference, since we know the animal, so to speak, only from the outside. We can only guess at the state of mind which impelled it to perform a certain act; and we anthropomorphize too much, when we attribute it to the same motive and prevision which would induce us to do likewise. The same cry or gesture may spring from very different emotions under different circumstances, when the power of expression is very limited in its range. Thus, the barking of a dog may be either a cry of alarm, a note of defiance, an outbreak of weariness, or an invitation to joyous frolic. In all these cases, the dog is merely giving voice indiscriminately to any strong emotion. The howling of the same animal is certainly a mournful sound, but it is not necessarily sadness or grief that makes it howl. The essence of language is the purpose or intention to communicate definite thought or emotion to others, and not the mere fact that the feeling is thus imparted, tho perhaps unintentionally. Through its inarticulate cries and gestures, the emotions of one animal may be made known to its fellows, and may thus actually spread an alarm among them, tho certainly not intended as a signal of danger, since the same cries are often repeated when the animal is alone, and there are none who can profit by the warning. If the imprisoned starling, that so much excited the sickly sensibility of Laurence Sterne[4] by its pitiful cry, "I can't get out," had been set at liberty by him, it would still have repeated the words as frequently as ever, and with quite as much perception of their actual meaning.

Let us attempt to clear the way for the consideration of this difficult question, by first enumerating the several powers and capacities which the lower animals unquestionably possess in common with man. We shall thus find ample grounds for our involuntary sympathy with many of them, and shall obtain a clearer understanding of the remaining points wherein their marked deficiencies consist. In the first place, there is no outward act whatever, considered simply as an exercise of nerves and muscles, which brutes cannot perform about as well as man, or even better. They leap, run, climb, and swim; they construct their homes, continue their species, and provide for their young; they weep, howl, and even articulate. Judging from the indications afforded by these outward acts, we have good reason to conclude that the senses of the lower animals, especially those of vision, hearing, and smell, are in many cases more acute and far-reaching than ours. It is equally evident that they have most of the emotions and passions, the desires and appetites, which incite and govern human conduct. They love, fear, and hate; they are angry, emulous, and revengeful. They are capable of magnanimity, and often clearly indicate curiosity, admiration, and *ennui.* Most of them are gregarious in their habits; they seek the company of their fellows, aid each other, are fond of sport, and distinguish their proper food. Their parental affections are very strong, leading almost to any amount of self-sacrifice, so long as their young need care and protection; but when this period of dependence is outgrown, they seem no longer even to recognize their offspring. They are also much under the influence of habit, and the imitative propensity appears clearly in many of them; the possibility of domesticating them, and training them to the service of man, depending largely upon the development of these two traits.

Passing to the more intellectual part of their nature, it is obvious that many of the brutes have a vivid imagination, and most of them have great facility in recognizing scenes and individuals of which they have had experience. A dog asleep upon the rug before the fire shows, by growls and barks, distinctly enough that he is stemming again in fantasy all the currents of a heady fight. Horses, dogs, and cats retrace with great precision a long road which they have but once travelled over, and never fail to stop at the gate or door which leads to their own home. The doctor's horse knows where most of the patients live about as well as the doctor himself. I do not here allude to any of the cases of supposed sagacity, foresight, and contrivance which are clearly attributable to instinct, an anomalous faculty with which all the lower animals are specially endowed; these are no indications of intellect properly so called, but are rather proofs of the absence of it, or of some serious defect of the mental constitution which needs to be supplemented by the action of so wonderful a substitute as this for reason in the strict sense of the term. Such cases will be considered hereafter, under their appropriate head as manifestations of instinct.

Animals are also capable of mental association, tho they certainly do not associate *ideas;* for, as will be shown hereafter, they have no ideas. They have only the immediate presentation through the senses of particular sights, sounds, odors, etc., together with the power of reviving these in imagination, of recognizing them as sensations formerly experienced, and of associating with them the emotions and passions with which they were originally accompanied. This is the simplest kind of association; it can all be resolved into an effect of habit, and is therefore closely akin to the animal's power of retracing the road which it has once travelled. The brute associates strongly the passion or emotion once aroused with the implement, the person, or the act by which it was first excited. It may even happen, as the animal is more excitable at one time than another, that the mere sight of the implement, or a menace of the repetition of the offensive act, will produce a greater burst of fury than resulted from the original infliction of pain. A dog will bear patiently a good deal of teasing, and a horse will submit to frequent touches with the whip, tho by quickening their pace, and trying to get out of the way, they show clearly enough that the act caused pain and resentment. But there comes a time when the brute is unusually sensitive, and some trivial annoyance, or a mere flourish of the whip, will excite a dangerous burst of passion. An invariably attendant sight or sound, harmless in itself, is also associated with the offensive act as its sign or symbol; and then the familiar cluck, chirrup, or whistle will produce the required act or emotion as surely as if accompanied with a blow. These facts not only explain the theory and method of training domestic animals, but explain away many supposed cases of sagacity, of acting on a premeditated plan, and of nursing suppressed resentment till a fit opportunity arrives for its full manifestation. Many such stories are told of the elephant, which one who has had large opportunities for observation has recently declared to be a remarkably stupid animal.

What has thus far been admitted undoubtedly tends, as far as it goes, to place the human and the brute mind on a par with each other. Is there, then, any strongly marked and unquestionable defect in the latter, which, when taken together with all its causes and consequences, places an impassable gulf between the two? Undoubtedly there is. The brute is utterly incapable of using language; it certainly cannot talk. This incapacity does not come from any defect in its physical organization, since parrots and several other birds may be taught to articulate words with great distinctness; and many of the mammalia can utter cries and make gestures which might, by convention, become as intelligible as the finger alphabet of the human deaf and dumb. Then their inability to talk must arise from the peculiarities of their mental constitution; and an analysis of the intellectual processes which are involved in the intelligent use of language will show clearly what are the inherent defects in the brute mind. There was as much argument as wit in the remark of a German naturalist, who said, "I will believe that animals have reason, when one of them shall tell me so." Good old John Locke[5] has been much ridiculed for merely citing a story, which he does not profess to believe, about a parrot owned by Prince Maurice, in Brazil, which was able to keep up an intelligent conversation with its visitors. Any attempt to teach animals to use language *meaningly* would deservedly excite equal ridicule, since their utter incapacity in this respect is obvious even to the vulgar. Laura Bridgman,[6] blind, deaf, and dumb from infancy, and thus apparently less fitted for communication with the external world than any of the vertebrate brutes, was yet mentally endowed with an innate capacity for the use of language, which has been so far developed by skilful instruction, that she now keeps a diary, and writes letters, with as large a use of significant phraseology as most educated persons possess. There never was a better illustration than her case presents of the etymological meaning of the word "education," that it signifies *bringing out* of the mind its native capacities, and not merely *putting into* it any amount of useful information. But no Dr. Howe has ever been foolish enough to attempt to teach a parrot or a monkey to converse, to write a significant sentence, or to read what is thus written. Balaam's ass did not rebuke its master except by a miracle.[7]

What must be the nature of that inherent mental defect which produces this absolute incapacity for the use of intelligible speech? The answer to this question is not doubtful or far to seek. Every significant word in any language is the name or symbol of a "concept;" that is, of what we English formerly called "an abstract general idea." As John Locke remarked long ago, brutes cannot generalize, and therefore they have no ideas to express, and cannot attach any meaning to words as uttered by others. As Bischoff[8] wittily puts it, "the plain reason why animals cannot talk is that they have nothing to say." They can be taught the symbolism of a few proper names; but proper names are not *words,* for they do not connote any meaning, and are therefore not susceptible of definition. Like a colored string hung round the neck, or a chalk-mark on the back, they *denote* the particular and individual act, man, or other object that is intended, but they do not *per se* connote any idea. If I previously know what message is to be imparted, the proper name, or the wave of a handkerchief, may point out to me the person to whom it is to be given. But such a name has no more significance in itself than the wave of the handkerchief. Every word properly so called is the name, not of an individual, but of a class, and it is applicable indifferently to all the members of the class, because it signifies (connotes) those qualities, and only those qualities, which are common to the whole class. Hence, comparison and discernment are needed, in order to know what individuals belong to the class; abstraction is required, so as to confine our attention to their common qualities; and generalization is the mental power whereby we recognize the commonness of those attributes, and the universality of the word within the limits of that class. Each of the mental acts here enumerated is an exercise of *judgment*, and the expression of judgments is the function of language. Hence the proper unit or fundamental element of significant speech is not the word, but the proposition, A single word, "man," does not express any cognition, or impart any fact or incident, except so far as it is explicated and made clear to thought by one or more of the judgments out of which it was first constituted, such as "man is mortal," "man is rational," etc. Consequently, brutes cannot grasp the meaning of a word, because they are mentally incapable of forming a judgment; *i.c.,* of *thinking a sentence*, wherein a predicate is significantly affirmed of a subject. If one of them hears a word of command, it is to him only a symbol, like a whistle or wave of the hand, with which is associated a certain emotion, and, through the unconscious force of habit, the emotion thus excited, leads him to perform the act which his master intended. It is of no use to utter a sentence or state a proposition to him, for he cannot understand it, and it will have no more effect on him than a single word, or catcall, or a gesture. The phrase "the human understanding" is a pleonasm, since every understanding is human or divine. The brute has no understanding, because it is incapable of *thought* strictly so called; that is, of comparison, discernment, and classification. Through the force of habit, and of associating emotions, as of pain and pleasure, with their signs, the animal is capable of being *trained;* but it is not susceptible of *education.* Nothing can be brought out of its mind, because nothing preexists in it which partakes of the nature of thought. Then the gulf between the brute and the human mind can never be bridged over; the two things being radically unlike, one might as well attempt to develop a football into a syllogism.

Because they have no thoughts of their own, and are incapable of interpreting the thoughts of others, the lower animals, as Schopenhauer[9] remarks, live entirely in the present. They have no proper past or future. Their mental horizon is strictly limited to the objects and events which *now* affect their senses. It is true, that they may imagine sensations which do not actually exist; but they cannot distinguish them as imagined from those which are real. The past may be presented in imagination, but it is not recognized as past; that is, it is not consciously assigned to a definite previous experience. It is merely a fictitious enlargement of the scene that is actually before the eyes. Because incapable of comparison and discernment in thought, the animal cannot distinguish the fictitious from the real, or what is, from what was, present to sense. Hence it probably does not apprehend either time or number, since these are not direct presentations of sense, but can only be cognized in thought. Indeed, time and number are mutually dependent; neither can be recognized without the other. Time can be thought only as a succession, a larger or smaller number, of moments; and number can be apprehended only by telling off successive units through a longer or shorter interval of time. Many familiar facts seem to indicate that the brutes have no sense of number. One puppy after another may be secretly abstracted from a numerous litter, and the mother shows no uneasiness or sense of loss; but she whines piteously after the last one is taken. The hen acts in a similar manner with a numerous brood of chickens. Hence, because incapable of numeration, the brute cannot distinguish between a longer and a shorter interval of time; that is, it has no apprehension of time as such. More briefly still, to distinguish the present from the past would require an act of comparison, which the animal has no power to perform. Being without a conscious past, the brute is also without a future, since even to human foresight, the future is only the shadow which the past throws in advance. The present moment, with its special sensations and emotions, its pleasures and its pains, either coming singly or associated by habit, forms the whole conscious life of the cat and the dog. They have no apprehension of the future, and therefore no dread of death.

[. . .]

A brute cannot be made wiser by any amount of experience because it cannot compare and distinguish, and is therefore incapable of apprehending any general truth, such as that which concerns the course of nature. It is always a slave of habit, whether the oft-repeated act is beneficial or injurious. It may be corrected by discipline, it is true, but only because a feeling of pain or pleasure is thereby artificially and blindly associated with an action otherwise meaningless.

Every one's observation of familiar facts will supply instances enough to prove this stationary character of the brute mind which arises from its inability to profit by experience. The moth flies again into the flame which had repeatedly singed its wings within a few minutes. A bee or wasp, attempting to escape from a room which it had accidentally entered, will knock its head fruitlessly for a long time against the same pane of glass, tho it might find free egress a few inches lower down. The typical form of nest, cell, or web, and the same routine of conduct, are blindly and persistently repeated, tho a change of circumstances has made them inadequate or useless for the end in view, and tho a slight and easy modification of them would render them again useful and agreeable. The act and the structure, which were needed and even indispensable while the animal was in its wild state, are renewed in its domesticated condition, tho they have then become meaningless and even injurious. The tamed squirrel, which has received more food than it craves for the moment, will scratch at the bottom of its wire cage and place a nut there, tho it should have learned from experience that it is no longer necessary to hide a store of food for future exigencies. In like manner, a half-domesticated beaver, which had the run of the house, attempted to build a dam with any materials that came in its way, tho there was no water near, and neither shelter nor concealment were now required, The lesson which nature originally taught the animal was so thoroughly learned that it is repeated by rote, come what may, evidently without any perception of its real meaning, or any ability to supplement it by the teachings of experience. Hence it is that the peculiar home, which every species of insect and bird constructs for itself, is built on the same pattern through an indefinite lapse of centuries, no improvements being copied from its neighbors, and none suggested by a forced change in its locality and modes of life.

This essential defect of the brute mind enables us to appreciate the breadth of the gulf which separates it from the human intellect, when we consider that man depends entirely upon experience for the preservation of his life, the fulfilment of his purposes, and the daily and even hourly regulation of his conduct. It is only by actual trial and through many efforts, failures, and errors, that we learn the rules of prudence, and how to find our way through the labyrinth of this world's affairs. It is only by experience that we are enabled to keep out of fire and water, to distinguish our food from our poison, to separate our friends from our enemies, and either to help others or to save ourselves. On this single foundation, indeed, is built up the whole fabric of human knowledge; for altho primitive convictions and truths spiritually discerned enter into the structure, and in some measure regulate its growth and determine its character, these *a priori* elements become pertinent and available for the guidance of conduct only so far as actual occurrences furnish occasions on which we may reduce them to practice. They supply, to adopt Kant's phraseology, only *the forms* of cognition, and these are comparatively abstract and void till our life's history furnishes *the matter* to which they are applicable. Conscience, for example, bids me observe certain principles of action, but leaves me to learn from observation how best to act upon them, and what are the tests of their due observance either by myself or others.

But each man's personal and individual experience is far too narrow and limited, especially in the earlier periods of life, to furnish adequate guidance for all exigencies that may arise, or a sufficient foundation for all the knowledge that he craves. It needs to be largely supplemented by other men's experience, whether these are our contemporaries or the members of former generations. Hence the peculiarly human endowment of language, as the indispensable means for the advancement of knowledge and the improvement of life, through the cumulation of the experience of the race. The lower animals have no need of communicating with each other, because they have no useful lessons to teach; as no one of them can put to any use the little store of his own experience, he would gain nothing by increasing this store through what might be added to it by his fellows. They cannot accumulate or transmit knowledge, because they are originally incapable of profiting by their own experience, and therefore they have nothing to impart to others. As already remarked, they do not talk because they have nothing to say.

We can now see a reason for the wonderful fact, to which I have already adverted – that, at least in the organic living world, nature takes good care never to exactly repeat herself. She never, even by accident, makes any one living thing indistinguishably like another. It is not that she is incapable of producing perfect uniformity in her work; for she does produce it in the inorganic world, where uniformity is the rule and any departure from it is the exception. The specific gravity of any elementary substance, the proportions in which such substances are chemically united into compounds, the definite forms into which they crystallize, the modes of action or affinities of different reagents, and many other instances of nature's work in this province, are precisely similar to each other; they do not vary even by a hair's breadth. Far otherwise is it in the world of living organisms, where variety is the rule and uniformity is the exception; nay, it is not even the exception, for not one such exception – that is, no case of two indiscernibles – can be produced. [. . .]

Now I say that this measureless variety of tint and outline, with which nature individualizes all her living products, is the necessary means of enabling experience to do its appropriate work. Only because every living man and animal is what the Scotch call *kenspeckle* – *i.e.,* easily recognized – can any one of us profit by the lessons of the past as a guide to the future. Thus only are we enabled to select appropriate means for definite purposes and ends. Thus only can we distinguish our friends from our enemies, him whom we may safely trust from him whom we must beware of, our food from our poison, my own child from a stranger to my blood, an explored country from a trackless waste. The whole fabric of civilized society may be said to depend upon the possibility of giving testimony on oath, that this particular man stole this particular horse. In other words, discrimination is necessary, and this, as we have seen, is the essential and distinctive element of human thought.

But the question immediately arises, What substitute does the brute possess for this peculiar endowment whereby the human mind is made capable of doing its necessary work, and thereby of providing for self-preservation and all the manifold exigencies of its life? Such a substitute there must be. Man is enabled to provide for his wants only because he can profit by his own experience and that of his fellows; while the brute, as we have seen, is not made one whit the wiser by any amount of experience, because it cannot properly distinguish one case from another, or discern the comparative faults and excellences of two of the same kind. How, then, can it preserve its life and perform its ordinary work? The answer to this question brings us at once to the heart of our subject, as it shows that the greatest difference between the human and the brute mind is not one of defect on the part of the latter, but rather of an excellence, since it is the exclusive possession of a faculty which so far transcends the power of human reason that, in its most common manifestations, it appears inscrutable, miraculous, and even divine. So far as the lower animals are guided by it – and they are all more or less under its control and dependent upon it – their actions appear to us not as subhuman, but as superhuman. Instinct is inspiration; even Kant says of it that "it is, the voice of God." The only adequate definition of Instinct is that it does all the work of experience without any aid from experience, so that man can only wonder at, but cannot understand, its operations.

It is a curious and instructive fact, that we have no one word in the English language to express what the French call *clairvoyance* and the Germans *Hellsehen;* that is, the assumed power of knowing more than experience ever taught, or is even capable of teaching. I say this is an instructive fact, because it indicates that practical and incredulous turn of the English mind, that large and roundabout common-sense, which recoils with aversion from all idle tales of seeing into the future without any aid from the past, or perceiving by immediate intuition what is obviously beyond the range of the senses, and generally of accomplishing any feat which transcends our ordinary human faculties. We have not the name, because we do not believe in the thing. The French, and especially the Germans, are more fond of the marvellous, and more prone to accept it on insufficient evidence. The Scotch give to the power in question the name of "second-sight." The nearest approach that we can make to this meaning in English is our word "seer," or prophet, denoting one who

"can look into the seeds of time,
And say which grain will grow and which will not."[10]

It is the power attributed by the vulgar to witches, demoniacs, somnambulists, and magnetized persons,[11] but which no educated and well-balanced mind is capable of believing that they actually possess.

Now I affirm that this very marvellous faculty of *Hellsehen* or *clairvoyance*, which no sensible person believes for a moment that any human mind has ever manifested, at least for the last eighteen centuries, is unquestionably possessed in high perfection by many, if not most, of the lower animals, under the name of Instinct, and even by those who are as low down in the scale as the fishes and the insects. I might take as a familiar instance the often-cited case of a kind of wasp, which stores up food of a sort which it never uses for itself, for the future sustenance of its young, whom it is never to see, because its own life ends before theirs begins. This case is instructive, because such prevision, confessedly beyond the range of the animal's individual experience, seems also to be inexplicable even by the aid of Mr. Herbert Spencer's ingenious supplemental theory of the accumulated effects of transmitted ancestral experience. But as the heroic imaginations of the evolutionists, which are not to be appalled by many difficulties so long as an opening can be forced through them by any suppositions however violent, do attempt to explain away such wonders by the combined influence of heredity and natural selection – as Mr. Darwin does, for instance, in respect to the transmitted instincts of the working neuter bees – I will take another case of instinct, which is undoubtedly independent both of individual and ancestral experience. I refer to the marvellous power, which many animals unquestionably possess, of finding their way home or to their proper point of migration, unaided by their previous explorations. The stories seem to be well authenticated as to the possession of this remarkable instinct even by mammals, such as the dog, the cat, and the donkey, which have been carried away by sea for a great distance, and have then found their way home overland, unguided, by a route never before traversed by them or their progenitors.

[. . .]

In fact, every undoubted case of instinct involves the exercise of this mysterious power of *Hellsehen*, or "second-sight for its nature is to make laborious provision for exigencies that are still in the comparatively remote future, and of which the animal has had no experience whatever. The bird not yet a year old, how does it know that the cares of maternity are coming upon it, and must be met by the construction of a nest of which each species has its distinctive pattern? Properly speaking, of course, the animal does not see into the future, its vision being strictly limited to what is before it at the present moment. But its actions and endeavors are as wisely regulated as if it had far more than human foresight. Instinctive action is working for a purpose without any consciousness of that purpose. That purpose is an all-important one, either for the continuance of the animal's own life or the propagation of its species. The long-continued and laborious work that is done for this purpose is usually distasteful for the moment, involving a considerable sacrifice of the creature's present ease and enjoyment. But urged by an impulse stronger than death, it lifts the self-imposed burden and bears it stoutly to the end, uncheered even by that which is the great solace of all human labor, the hope of future happiness which is to reward the performance of duty. The action of instinct bears some resemblance to what man does under the influence of habit, his work having become a mere routine, which he performs without reflection and almost without consciousness. But habit is not instinct, for it is slowly generated and perfected by experience; while instinct, perfect from the moment of the animal's birth, is altogether independent of experience.

Instinct is certainly given to the brutes as a substitute both for reason and for experience which supplies the materials on which human reason operates; and, in truth, we know that the two faculties exist in inverse ratio to each other. As at the bottom of the scale, in the lowest animal, there is certainly no trace of reason, so at the top, in man, there is no vestige of instinct, In the human being, it is true, we find natural and primitive emotions and appetites, which are often loosely called "instinctive." But they do not deserve the name, for they dictate only the end to be pursued, but do not guide us, as instinct would, in selecting the right means for its attainment. On the contrary, the stronger the feeling or desire, too frequently are we the more mistaken in our eager attempts to gratify it, which often defeat the very purpose we have in view. Instinct does not commit such blunders. I have already alluded to the fact, that the highest and most marvellous instincts are manifested far down in the scale, among ants, spiders, and bees, and are comparatively infrequent among the vertebrates. But even here birds show more numerous and more complex instincts than mammals, and fishes are guided more by what answers to inspiration than by habit. Hence, as Prof. Mivart[12] remarks, the more instinctive a man's actions are, the less are they rational, and *vice versa;* and this amounts to a demonstration that reason has not, and by no possibility could have, been developed from instinct. "When two faculties tend to increase in inverse ratio, it becomes unquestionable that the difference between them is one of kind." [. . .]

The instinct which guides the ant and the bee is really the same with the power or agency, call it what you may, which directs the physiological processes that paint the peacock's tail with so complex and gorgeous a pattern, and arrays shells and flowers in their decorated holiday garb. Indeed, the process of development through which the whole organism is built up on a complex but definite plan, with its machinery of limbs, muscles, joints, and nerves, and its adornment of plumage, shell, and scales, is but half of the work necessary to be accomplished before the animal is fitted to play the part assigned to it in nature's scheme, and thus to preserve its own life and continue its species. Together with the organs and other physical means of doing its work, it must have the knowledge and skill requisite for making the proper use of those means, and directing them towards the appropriate ends to be accomplished. The instinct and the organism in which it is lodged are thus necessarily related to each other as parts of one whole; any change in either of the two factors would incapacitate the animal for its task, if it were not accompanied by a corresponding change, nicely adapted to it, in the other. Thus, in the spider, the gland which secretes the viscid fluid, which is the raw material for constructing the web, must be correlated with the instinct for drawing out the threads and weaving them into the peculiar pattern of that web; and both these processes, again, must be nicely correlated to the general purpose in view, that of entrapping its prey.

Webbed feet and plumage impervious to wet must be correlated with the instinct to take to the water and swim; as is clearly seen in the case of ducklings that have been hatched out by a hen, as in this case, the instinct cannot have been acquired by instruction, experience, or imitation. Many birds instinctively hide their nests both by building them in crevices and corners not exposed to open view, and by assimilating them to the color of the surrounding rock or foliage. In like manner, the agency, whatever it may be, which constructs the animal's organism often provides for its protection through concealment by similar means, as by assimilating the color of its skin or plumage to that of its surroundings, or by mimicry of forms of a different nature; for instance, by imitating the color and external structure of a dead leaf. Even man has an involuntary and almost unconscious impulse to imitate actions in which he is much interested, as when we cough or yawn by contagion, or writhe and twist our bodies in sympathy with the rope-dancer whom we are gazing at. Reasoning from analogy, then, we may well conclude that it is an imitative impulse, tho a wholly unconscious one, which gradually assimilates the insect's color and external form to a dead leaf, or some other shape seemingly so fantastic that it appears like a purposeless freak of nature. But then we must also admit that the ordinary process of building up the animal's whole body in the normal way, after the common type of its species, is throughout the animal's own act, tho a blind and unconscious one, performed under the guidance of a higher power. It is not the brain which generates the instinct, but it is the instinct which constructs the brain, as well as every other portion of the organism.

FRANCIS BOWEN.

Editorial Endnotes

1 *Montaigne . . . followed by Charron and Gassendi*: this refers, respectively, to French Renaissance philosopher Michel de Montaigne (1533–1592), French theologian Pierre Charron (1541–1603), and French philosopher Pierre Gassendi (1592–1655).

2 *Herbert Spencer*: Herbert Spencer (1820–1903) was a highly influential British philosopher, sociologist, and evolutionary theorist whose works were widely discussed in the United States.

3 *Mr. Huxley*: Thomas Henry Huxley (1825–1895) was an English biologist and educator. He was a leading proponent and populariser of Darwinian evolutionary theory, earning him the nickname "Darwin's bulldog."

4 *Laurence Sterne*: Laurence Sterne (1713–1768) was an Anglo-Irish novelist mostly known today for his fictional autobiography *The Life and Opinions of Tristram Shandy, Gentleman*, first published in nine volumes between 1759 and 1767.

5 *John Locke*: John Locke: (1632–1704), an English philosopher who played a significant role in the development of modern political theory – especially with his ideas about natural rights, the social contract, and government by consent – and was a major influence on Enlightenment thought.

6 *Laura Bridgman*: Laura Bridgman (1829–1889) was the first deaf-blind person to receive an extensive formal education and successfully communicate in sign language. From the age of eight, she was taught at the Perkins Institution for the Blind in Massachusetts by the institution's first director, Samuel Gridley Howe (also mentioned in the article).

7 *Balaam's ass did not rebuke its master except by a miracle*: a story from the Book of Numbers. Balaam is a diviner sent to curse the Israelites at the behest of King Balak, ruler of Moab (in the area of modern-day Jordan), who wants to protect his kingdom from the new arrivals. An angel is sent to Balaam but is only noticed by his riding animal, a donkey, who is eventually given the ability to speak to Balaam, revealing the angel's presence to him. The scene is captured in Dutch artist Pieter Lastman's *Balaam and the Ass* (1622) and the 1626 reiteration of the painting by Lastman's famous student Rembrandt.

8 *Bischoff*: likely Theodor Ludwig Wilhelm von Bischoff (1807–1882), a German biologist and physiologist. However, variations of this saying have been used by, or attributed to, a number of authors, including the philosophers Wilhelm von Humboldt and Wilhelm Wundt.

9 *Schopenhauer*: Arthur Schopenhauer (1788–1860) was a German pessimist philosopher. Animal consciousness and the moral treatment of animals are recurring topics in his philosophy, including in his principal work *The World as Will and Representation* (*Die Welt als Wille und Vorstellung*, first published in 1818).

10 *"can look into . . . will not"*: spoken by Banquo in Act I, Scene 3 of Shakespeare's *Macbeth* (likely written in 1606), when he asks the Three Witches to tell him the future.

11 *magnetized persons*: this is a reference to the notion of "animal magnetism" or "mesmerism" (after Vienna physician Franz Anton Mesmer) and the controversial therapy associated with it. Mesmerism posited the existence of a mysterious "magnetic fluid" whose flow could be controlled by "magnetisers" to induce trance-like states and achieve therapeutic benefits or even cure diseases.

12 *Prof. Mivart*: Saint George Mivart (1827–1900) was an English biologist and philosopher known for his criticisms of Darwinian natural selection. In 1881, Mivart published a well-received book on cat anatomy titled *The Cat: An Introduction to the Study of Backboned Animals*.

Editorial Headnote

7. William James, "What Is an Instinct?" *Scribner's Magazine* 1, no. 3 (March 1887): 355–65.

Harvard philosopher and psychologist William James (1842–1910) is considered a founding figure of the philosophical school of pragmatism, which eschewed abstract metaphysical debates and theorisation in favor of a general emphasis on the practical value and problem-solving qualities of knowledge (production). James's two-volume *Principles of Psychology* (1890), in which he developed his theoretical perspectives on (the stream of) consciousness, habit, emotion, and other phenomena, is considered a landmark work of modern psychology and quickly became the rising discipline's principal textbook in the United States and Europe.[1]

As a Darwinian, James looked at psychological phenomena through the lens of evolutionary theory (earning him the opposition of Bowen, the university's most prominent philosopher during James's early years at Harvard), which included the argument that animal instincts were the result of natural selection. In the following article for *Scribner's Magazine*, an illustrated monthly launched by the sons of New York publisher Charles Scribner I to compete with heavyweights like *Harper's* that ran from 1887 to 1939,[2] James argues against the commonplace notion that the difference between human and animal behavior boiled down to the difference between reason and instinct and against the concomitant idea that the significance of the one came at the expense of the other. While this idea implied that humans, as reasoning beings *par excellence*, could be understood as creatures of instinct only in a very limited sense, James argues that the opposite was true: it was a *proliferation* rather than a poverty of instincts that defined human beings, and human beings were no less governed by such instincts or "impulses" than other animals. For James, the reason why humans seemed to be uniquely rational and deliberative in their actions was because, first, in their multiplicity human instincts competed with and inhibited each other and, second, instincts exercised their full dispositional force only upon their first appearance and were afterward modified by our experience of them, thus losing much of their rigidity and predictability. While this "masking of the elementary condition of the instinctive life" was particularly significant in humans, it was by no means exclusive to them: other animals, too – especially "higher" ones – were influenced in their behavior by competing or contradictory instincts, nor were they doomed to "blindly" exercise the same invariable reflex behavior.

Notes

1 Ignas K. Skrupskelis, "James, William," in *American National Biography*, ed. John A. Garraty and Mark C. Carnes, vol. 11 (New York: Oxford University Press, 1999), 840–46.

2 On *Scribner's Magazine*, see Frank Luther Mott, *A History of American Magazines*, vol. 4: 1885–1905 (Cambridge: Harvard University Press, 1957), 717–32.

7

WHAT IS AN INSTINCT? (1887)

William James

Source: William James, "What Is an Instinct?" *Scribner's Magazine* 1, no. 3 (March 1887): 355–65.

INSTINCT is usually defined as the faculty of acting in such a way as to produce certain ends, without foresight of the ends, and without previous education in the performance. That instincts, as thus defined, exist on an enormous scale in the animal kingdom needs no proof. They are the functional correlatives of structure. With the presence of a certain organ goes, one may say, almost always a native aptitude for its use. [. . .]

A very common way of talking about these admirably definite tendencies to act is by naming abstractly the purpose they subserve, such as self-preservation, or defence, or care for eggs and young – and saying the animal has an instinctive fear of death or love of life, or that she has an instinct of self-preservation, or an instinct of maternity and the like. But this represents the animal as obeying abstractions which, not once in a million cases is it possible it can have framed. The strict physiological way of interpreting the facts leads to far clearer results. The actions we call instinctive all conform to the general reflex type; they are called forth by determinate sensory stimuli in contact with the animal's body, or at a distance in his environment. The cat runs after the mouse, runs or shows fight before the dog, avoids falling from walls and trees, shuns fire and water, etc., not because he has any notion either of life or of death, or of self, or of preservation. He has probably attained to no one of these conceptions in such a way as to react definitely upon it. He acts in each case separately, and simply because he cannot help it; being so framed that when that particular running thing called a mouse appears in his field of vision he *must* pursue; that when that particular barking and obstreperous thing called a dog appears there he *must* retire, if at a distance, and scratch if close by; that he *must* withdraw his feet from water and his face from flame, etc. His nervous system is to a great extent a preorganized bundle of such reactions – they are as fatal as sneezing, and as exactly correlated to their special excitants as it is to its own. Although the naturalist may, for his own convenience, class these reactions under general heads, he must not forget that

DOI: 10.4324/9781003032717-8

in the animal it is a particular sensation or perception or image which calls them forth.

At first this view astounds us by the enormous number of special adjustments it supposes animals to possess ready-made in anticipation of the outer things among which they are to dwell. *Can* mutual dependence be so intricate and go so far? Is each thing born fitted to particular other things, and to them exclusively, as locks are fitted to their keys? Undoubtedly, this must be believed to be so. Each nook and cranny of creation, down to our very skin and entrails, has its living inhabitants, with organs suited to the place, to devour and digest the food it harbors and to meet the dangers it conceals; and the minuteness of adaptation, thus shown in the way of *structure,* knows no bounds. Even so are there no bounds to the minuteness of adaptation in the way of *conduct* which the several inhabitants display.

The older writings on instinct are ineffectual wastes of words, because their authors never came down to this definite and simple point of view, but smothered everything in vague wonder at the clairvoyant and prophetic power of the animals – so superior to anything in man – and at the beneficence of God in endowing them with such a gift. But God's beneficence endows them, first of all, with a nervous system; and, turning our attention to this, makes instinct immediately appear neither more nor less wonderful than all the other facts of life.

Every instinct is an *impulse.* Whether we shall call such impulses as blushing, sneezing, coughing, smiling, or dodging, or keeping time to music, instincts or not, is a mere matter of terminology. The process is the same throughout. In his delightfully fresh and interesting work, "Der Thierische Wille," Herr G. H. Schneider[1] subdivides impulses (*Triebe*) into sensation-impulses, perception-impulses, and idea-impulses. To crouch from cold is a sensation-impulse; to turn and follow, if we see people running one way, is a perception-impulse; to cast about for cover, if it begins to blow and rain, is an imagination-impulse. A single complex instinctive action may involve successively the awakening of impulses of all three classes. Thus a hungry lion starts to *seek* prey by the awakening in him of imagination coupled with desire; he begins to *stalk* it when, on eye, ear, or nostril, he gets an impression of its presence at a certain distance; he *springs* upon it, either when the booty takes alarm and flees, or when the distance is sufficiently reduced; he proceeds to *tear* and *devour* it the moment he gets a sensation of its contact with his claws and fangs. Seeking, stalking, springing, and devouring are just so many different kinds of muscular contraction, and neither kind is called forth by the stimulus appropriate to the other. Schneider says of the hamster, which stores corn in its hole: "If we analyze the propensity of storing, we find that it consists of three impulses: First, an impulse to *pick up* the nutritious object, due to perception; second, an impulse to *carry it off* into the dwelling-place, due to the idea of this latter; and third, an impulse to *lay it down* there, due to the sight of the place. It lies in the nature of the hamster that it should never see a full ear of corn without feeling a desire to strip it; it lies in its nature to feel, as soon as its cheek-pouches are filled, an irresistible desire to hurry to its home; and finally, it lies it its nature that the sight of the storehouse should awaken the impulse to empty the cheeks" (p. 208). In certain animals of a low order the feeling of having executed one impulsive step is such an indispensable part of the stimulus of the next one, that the animal cannot make any variation in the order of its performance.

Now, why do the various animals do what seem to us such strange things, in the presence of such outlandish stimuli? Why does the hen, for example, submit herself to the tedium of incubating such a fearfully uninteresting set of objects as a nestful of eggs, unless she have some sort of a prophetic inkling of the result? The only answer is *ad hominem.* We can only interpret the instincts of brutes by what we know of instincts in ourselves. Why do men always lie down, when they can, on soft beds rather than on hard floors? Why do they sit round the stove on a cold day? Why, in a room, do they place themselves, ninety-nine times out of a hundred, with their faces toward its middle rather than to the wall? Why do they prefer saddle of mutton and champagne to hard-tack and ditchwater? Why does the maiden interest the youth so that everything about her seems more important and significant than anything else in the world? Nothing more can be said than that these are human ways, and that every creature *likes* its own ways, and takes to the following them as a matter of course. Science may come and consider these ways, and find that most of them are useful. But it is not for the sake of their utility that they are followed, but because at the moment of following them we feel that that is the only appropriate and natural thing to do. Not one man in a billion, when taking his dinner ever thinks of utility. He eats because the food tastes good and makes him want more. If you ask him *why* he should want to eat more of what tastes like that, instead of revering you as a philosopher he will probably laugh at you for a fool. The connection between the savory sensation and the act it awakens is for him absolute and *selbstverständlich,* an "*a priori* synthesis"[2] of the most perfect sort, needing no proof but its own evidence. It takes, in short, what Berkeley[3] calls a mind debauched by learning to carry the process of making the natural seem strange so far as to ask for the *why* of any instinctive human act. To the metaphysician alone can such questions occur as: Why do we smile, when pleased, and not scowl? Why are we unable to talk to a crowd as we talk to a single friend? Why does a particular maiden turn our wits so upside down? The common man can only say, "*of course* we smile, *of course* our heart palpitates at the sight of the crowd, *of course* we love the maiden, that beautiful soul clad in that perfect form, so palpably and flagrantly made from all eternity to be loved!"

And so probably does each animal feel about the particular things it tends to do in presence of particular objects. They, too, are *a priori* syntheses. To the lion it is the lioness which is made to be loved; to the bear, the she-bear. To the broody hen the notion would probably seem monstrous that there should be a creature in the world to whom a nestful of eggs was not the utterly fascinating and precious and never-to-be-too-much-sat-upon object which it is to her.[a]

Thus we may be sure that, however mysterious some animals' instincts may appear to us, our instincts will appear no less mysterious to them. And we may conclude that, to the animal which obeys it, every impulse and every step of every instinct shines with its own sufficient light, and seems at the moment the only eternally right and proper thing to do. It is done for its own sake exclusively. What voluptuous thrill may not shake a fly, when she at last discovers the one particular leaf, or carrion, or bit of dung, that out of all the world can stimulate her ovipositor to its discharge? Does not the discharge then seem to her the only fitting thing? And need she care or know anything about the future maggot and its food?

[. . .]

Remember that nothing is said yet of the origin of instincts, but only of the constitution of those that exist fully formed. How stands it with the instincts of mankind?

Nothing is commoner than the remark that Man differs from lower creatures by the almost total absence of instincts, and the assumption of their work in him by "reason." A fruitless discussion might be waged on this point by two theorizers who were careful not to define their terms. "Reason" might be used, as it often has been used since Kant, not as the mere power of "inferring," but also as a name for the *tendency to obey impulses* of a certain lofty sort, such as duty, or universal ends. And "instinct" might have its significance so broadened as to cover all impulses whatever, even the impulse to act from the idea of a distant fact, as well as the impulse to act from a present sensation. Were the word instinct used in this broad way, it would of course be impossible to restrict it, as we began by doing, to actions done with no prevision of an end. We must of course avoid a quarrel about words, and the facts of the case are really tolerably plain! Man has a far greater variety of *impulses* than any lower animal; and any one of these impulses, taken in itself, is as "blind" as the lowest instinct can be; but, owing to man's memory, power of reflection, and power of inference, they come each one to be felt by him, after he has once yielded to them and experienced their results, in connection with a *foresight* of those results. In this condition an impulse acted out may be said to be acted out, in part at least, *for the sake* of its results. It is obvious that every instinctive act, in an animal with memory, must cease to be "blind" after being once repeated, and must be accompanied with foresight of its "end" just so far as that end may have fallen under the animal's cognizance. An insect that lays her eggs in a place where she never sees them hatch must always do so "blindly;" but a hen who has already hatched a brood can hardly be assumed to sit with perfect "blindness" on her second nest. Some expectation of consequences must in every case like this be aroused; and this expectation, according as it is that of something desired or of something disliked, must necessarily either re-enforce or inhibit the mere impulse. The hen's idea of the chickens would probably encourage her to sit; a rat's memory, on the other hand, of a former escape from a trap would neutralize his impulse to take bait from anything that reminded him of that trap. If a boy sees a fat hopping-toad, he probably has incontinently an impulse (especially if with other boys) to smash the creature with a stone, which impulse we may suppose him blindly to obey. But something in the expression of the dying toad's clasped hands suggests the meanness of the act, or reminds him of sayings he has heard about the sufferings of animals being like his own; so that, when next he is tempted *by a toad,* an idea arises which, far from spurring him again to the torment, prompts kindly actions, and may even make him the toad's champion against less reflecting boys.

It is plain then that, no matter how well endowed an animal may originally be in the way of instincts, his resultant *actions* will be much modified if the instincts *combine* with experience, if in addition to impulses he have memories, associations, inferences, and expectations, on any considerable scale. An object O, on which he has an instinctive impulse to react in the manner A, would *directly* provoke him to that reaction. But O has meantime become for him a *sign* of the nearness of P, on which he has an equally strong impulse to react in the manner B, quite unlike A. So that when he meets O the immediate impulse A and the remote impulse B struggle in his breast for the mastery. The fatality and uniformity said to be characteristic of instinctive actions are so little manifest, that one might be tempted to deny to him altogether the possession of any instinct about the object O. Yet how false this judgment would be! The instinct about O is there; only by the complication of the mental machinery it has come into conflict with another instinct about P.

Here we immediately reap the good fruits of our simple physiological conception of what an instinct is. If it be a mere excito-motor impulse, due to the preexistence of a certain "reflex-arc" in the nerve-centres of the creature, of course it must follow the law of all such reflex-arcs. One liability of such arcs is to have their activity "inhibited" by other processes going on at the same time. It makes no difference whether the arc be organized at birth, or ripen spontaneously later, or be due to acquired habit, it must take its chances with all the other arcs, and sometimes succeed, and sometimes fail, in draughting off the currents through itself. The mystical view of an instinct would make it invariable. The physiological view would require it to show occasional irregularities in any animal in whom the number of separate instincts, and the possible entrance of the same stimulus into several of them, were great. And such irregularities are what every superior animal's instincts do show in abundance.[b]

Wherever the mind is elevated enough to discriminate; wherever several distinct sensory elements must combine to discharge the reflex-arc; wherever, instead of plumping into action instantly at the first rough intimation of what *sort* of a thing is there, the agent waits to see which *one* of its kind it is and what the *circumstances* are of its appearance; wherever different individuals and different circumstances can impel him in different ways; wherever these are the conditions – we have a masking of the elementary constitution of the instinctive life. The whole story of our dealings with the lower wild animals is the history of our taking advantage of the way in which they judge of everything by its mere label, as it were, so as to ensnare or kill them. Nature, in them, has left matters in this rough way, and made them act *always* in the manner which would be *oftenest* right. There are more worms unattached to hooks than impaled upon them; therefore, on the whole, says Nature to her fishy children, bite at *every* worm and take your chances. But as her children get higher, and their lives more precious, she reduces the risks. Since what seems to be the same object may be now a genuine food and now a bait; since in gregarious species each individual may prove to be either the friend or the rival, according to the circumstances, of another; since any entirely unknown object may be fraught with weal or woe, Nature implants *contrary* impulses to act on many classes of things, and leaves it to slight alterations in the conditions of the individual case to decide which impulse shall carry the day. Thus, greediness and suspicion, curiosity and timidity, coyness and desire, bashfulness and vanity, sociability and pugnacity, seem to shoot over into each other as quickly, and to remain in as unstable equilibrium in the higher birds and mammals as in man. They are all impulses, congenital, blind at first, and productive of motor reactions of a rigorously determinate sort. Each one of them, then, is an instinct, as instincts are commonly defined. But they contradict each other – "experience" in each particular opportunity of application usually deciding the issue. The animal that exhibits them loses the "instinctive" demeanor and appears to lead a life of hesitation and choice, an intellectual life; not, however, because he has no instincts – rather because he has so many that they block each other's path.

Thus, then, without troubling ourselves about the words instinct and reason, we may confidently say that however uncertain man's reactions upon his environment may sometimes seem in comparison with those of lower creatures, the uncertainty is probably not due to their possession of any principles of action which he lacks, but to his possessing all the impulses that they have, and a great many more besides. In other words, there is no material antagonism between instinct and reason. Reason, *per se,* can inhibit no impulses; the only thing that can neutralize an impulse is an impulse the other way. Reason may, however, make an *inference which will set loose* the impulse the other way; and thus, though the animal richest in reason might be also the animal richest in instinctive impulses too, he would never seem the fatal automaton which a *merely* instinctive animal would be.

Let us now turn to human impulses with a little more detail. All we have ascertained so far is that impulses of an originally instinctive character may exist, and yet not betray themselves by automatic fatality of conduct. But in man what impulses do exist? In the light of what has been said, it is obvious that an existing impulse may not always be superficially apparent even when its object is there. And we shall see that some impulses may be masked by causes of which we have not yet spoken.

Were one devising an abstract scheme, nothing would be easier than to discover from an animal's actions just how many instincts he possessed. He would react in one way only upon each class of objects with which his life had to deal; he would react in identically the same way upon every specimen of a class; and he would react invariably during his whole life. There would be no gaps among his instincts; all would come to light without perversion or disguise. But there are no such abstract animals, and nowhere does the instinctive life display itself in such a way. Not only, as we have seen, may objects of the same class arouse reactions of opposite sorts in consequence of slight changes in the circumstances in the individual object or in the agent's inward condition; but two other principles, of which we have not yet spoken, may come into play and produce results so striking that observers as eminent as Messrs. D. A. Spalding and Romanes[4] do not hesitate to call them "derangements of the mental constitution," and to conclude that the instinctive machinery has got out of gear.

These principles are those of the *inhibition of instincts by habits* and of the *transitoriness of instincts.* Taken in conjunction with the two former principles – that an object may excite ambiguous impulses, or *suggest* an impulse different from that which it *excites,* by suggesting a remote object – they explain any amount of departure from uniformity of conduct, without implying any getting out of gear of the elementary impulses from which the conduct flows.

Take first the inhibition of instincts by habits. The law is this: When objects of a certain class elicit from an animal a certain sort of reaction, it often happens that the animal becomes *partial* to the first specimen of the class on which it has reacted, and will not afterward react on any other specimen.

The selection of a particular hole to live in, of a particular mate, of a particular feeding-ground, a particular variety of diet, a particular anything, in short, out of a possible multitude, is a very wide-spread tendency among animals, even those low down in the scale. The limpet will return to the same sticking-place in its rock, and the lobster to its favorite nook on the sea-bottom. The rabbit will deposit its dung in the same corner; the bird makes its nest on the same bough. But each of these preferences carries with it an insensibility to *other* opportunities and occasions – an insensibility which can only be described physiologically as an inhibition of new impulses by the habit of old ones already formed. The possession of homes and wives of our own makes us strangely insensible to the charms of those of other people. Few of us are adventurous in the matter of food; in fact, most of us think there is something disgusting in a bill of fare to which we are unused. Strangers, we are apt to think, cannot be worth knowing, especially if they come from distant cities, etc. The original impulse which got us homes, wives, dietaries, and friends, at all, seems to exhaust itself in its first achievements and to leave no surplus energy for reacting on new cases. And so it comes about that, witnessing this torpor, an observer of mankind might say that no *instinctive* propensity toward certain objects existed at all. It existed, but it existed *miscellaneously,* or as an instinct pure and simple only, before habit was formed. A habit, once grafted on an instinctive tendency, restricts the range of the tendency itself, and keeps us from reacting on any but the habitual objects, although other objects might just as well have been chosen had they been the first comers.

Another sort of arrest of instinct by habit is where the same class of objects awakens contrary instinctive impulses. Here the impulse first followed toward a given individual of the class is apt to keep him from ever awakening the opposite impulse in us. In fact, the whole class may be protected by this individual specimen from the application to it of the other impulse. Animals, for example, awaken in a child the opposite impulses of fearing and fondling. But if a child, in his first attempts to pat a dog, gets snapped at or bitten, so that the impulse of fear is strongly aroused, if may be that for years to come no dog will excite in him the impulse to fondle again. On the other hand, the greatest natural enemies, if carefully introduced to each other when young and guided at the outset by superior authority, settle down into those "happy families" of friends which we see in our menageries. Young animals, immediately after birth, have no instinct of fear, but show their dependence by allowing themselves to be freely handled. Later, however, they grow "wild," and, if left to themselves, will not let man approach them. I am told by farmers in the Adirondack wilderness that it is a very serious matter if a cow wanders off and calves in the woods and is not found for a week or more. The calf, by that time, is as wild and almost as fleet as a deer, and hard to capture without violence. But calves rarely show any particular wildness to the men who have been in contact with them during the first days of their life, when the instinct to attach themselves is uppermost, nor do they dread strangers as they would if brought up wild.

Chickens give a curious illustration of the same law. Mr. Spalding's wonderful article on instinct shall supply us with the facts. These little creatures show opposite instincts of attachment and fear, either of which maybe aroused by the same object, man. If a chick is born in the absence of the hen, it "will follow any moving object. And, when guided by sight alone, they seem to have no more disposition to follow a hen than to follow a duck or a human being. Unreflecting lookers-on, when they saw chickens a day old running after me," says Mr. Spalding, "and older ones following me for miles, and answering to my whistle, imagined that I must have some occult power over the creatures: whereas I had simply allowed them to follow me from the first. There is the instinct to follow; and the ear, prior to experience, attaches them to the right object."[c] But if a man presents himself for the first time when the instinct of *fear* is strong, the phenomena are altogether reversed. Mr. Spalding kept three chickens hooded until they were nearly four days old, and thus describes their behavior:

> "Each of them, on being unhooded, evinced the greatest terror to me, dashing off in the opposite direction whenever I sought to approach it. The table on which they were unhooded stood before a window, and each in its turn beat against the window like a wild bird. One of them darted behind some books, and, squeezing itself into a corner, remained cowering for a length of time. We might guess at the meaning of this strange and exceptional wildness; but the odd fact is enough for my present purpose. Whatever might have been the meaning of this marked change in their mental constitution – had they been unhooded on the previous day they would have run to me instead of from me – it could not have been the effect of experience; it must have resulted wholly from changes in their own organizations."[d]

Their case was precisely analogous to that of the Adirondack calves. The two opposite instincts relative to the same object ripen in succession. If the first one engenders a habit, that habit will inhibit the application of the second instinct to that object. All animals are tame during some phase of their infancy. Habits formed then limit the effects of whatever instincts of wildness may later be evolved.

Mr. Romanes gives some very curious examples of the way in which instinctive tendencies may be altered by the habits to which their first "objects" have given rise. The cases are a little more complicated than those mentioned in the text, inasmuch as the object reacted on not only starts a habit which inhibits other kinds of impulse toward it (although such other kinds might be natural), but even modifies by its own peculiar conduct the constitution of the impulse it actually awakens.

Two of the instances in question are those of hens who hatched out broods of chicks after having (in three previous years) hatched ducks. They strove to coax or to compel their new progeny to enter the water, and seemed much perplexed at their unwillingness. Another hen adopted a brood of young ferrets which, having lost their mother, were put under her. During all the time they were left with her she had to sit on the nest, for they could not wander like young chicks. She obeyed their hoarse growling as she would have obeyed her chickens' peep. She combed out their hair with her bill, and "used frequently to stop and look with one eye at the wriggling nestful, with an inquiring gaze, expressive of astonishment." At other times she would fly up with a loud scream, doubtless because the orphans had nipped her in their search for teats. Finally, a Brahma hen nursed a young peacock during the enormous period of *eighteen months,* and never laid any eggs during all this time. The abnormal degree of pride which she showed in her wonderful chicken is described by Dr. Romanes as ludicrous.[e]

This leads us to the law of transitoriness, which is this: That *many instincts ripen at a certain age and then fade away.* A consequence of this law is that if, during the time of such an instinct's vivacity, objects adequate to arouse it are met with, a *habit* of acting on them is formed, which remains when the original instinct has passed away; but that if no such objects are met with, then no habit will be formed; and, later on in life, when the animal meets the objects, he will altogether fail to react, as at the earlier epoch he would instinctively have done.

No doubt such a law is restricted. Some instincts are far less transient than others – those connected with feeding and "self-preservation" may hardly be transient at all, and some, after fading out for a time, recur as strong as ever, *e.g.,* the instincts of pairing and rearing young. The law, however, though not absolute, is certainly very wide-spread, and a few examples will illustrate just what it means.

In the chickens and calves above mentioned, it is obvious that the instinct to follow and become attached fades out after a few days, and that the instinct of flight then takes its place, the conduct of the creature toward man being decided by the formation or non-formation of a certain habit during those days. The transiency of the chicken's instinct to follow is also proved by its conduct toward the hen. Mr. Spalding kept some chickens shut up till they were comparatively old, and, speaking of these, he says: "A chicken that has not heard the call of the mother till until eight or ten days old then hears it as if it heard it not. I regret to find that on this point my notes are not so full as I could wish, or as they might have been. There is, however, an account of one chicken that could not be returned to the mother when ten days old. The hen followed it and tried to entice it in every way; still, it continually left her and ran to the house or to any person of whom it caught sight. This it persisted in doing, though beaten back with a small branch dozens of times and, indeed, cruelly maltreated. It was also placed under the mother at night, but it again left her in the morning."

The instinct of sucking is ripe in all mammals at birth, and leads to that habit of taking the breast which, in the human infant, may be prolonged by daily exercise long beyond its usual term of a year or a year and a half. But the instinct itself is transient, in the sense that if, for any reason, the child be fed by spoon during the first few days of its life and not put to the breast, it may be no easy matter after that to make it suck at all. So of calves. If their mother die, or be dry, or refuse to let them suck for a day or two, so that they are fed by hand, it becomes hard to get them to suck at all when a new nurse is provided. The ease with which sucking creatures are weaned, by simply breaking the habit and giving them food in a new way, shows that the instinct, purely as such, must be entirely extinct.

Assuredly the simple fact that instincts are transient, and that the effect of later ones may be altered by the habits which earlier ones have left behind, is a far more philosophical explanation than the notion of an instinctive constitution vaguely "deranged" or "thrown out of gear."

I have observed a Scotch terrier, born on the floor of a stable in December, and transferred six weeks later to a carpeted house, make, when he was less than four months old, a very elaborate pretence of burying things, such as gloves, etc., which he had played with till he was tired. He scratched the carpet with his forefeet, dropped the object from his mouth upon the spot, and then scratched all about it (with both fore and hind feet, if I remember rightly), and finally went away and let it lie. Of course, the act was entirely useless. I saw him perform it at that age, some four or five times, and never again in his life. The conditions were not present to fix a habit which could last when the prompting instinct died away. But suppose meat instead of a glove, earth instead of a carpet, hunger pangs instead of a fresh supper a few hours later, and it is easy to see how this dog might have got into a habit of burying superfluous food, which might have lasted all his life. Who can swear that the strictly instinctive part of the food-burying propensity in the wild *Canidæ* may not be as short-lived as it was in this terrier?

[. . .]

Leaving lower animals aside, and turning to human instincts, we see the law of transiency corroborated on the widest scale by the alternation of different interests and passions as human life goes on. With the child, life is all play and fairy-tales and learning the external properties of "things;" with the youth, it is bodily exercises of a more systematic sort, novels of the real world, boon-fellowship and song, friendship and love, nature, travel and adventure, science and philosophy; with the man, ambition and policy, acquisitiveness, responsibility to others, and the selfish zest of the battle of life. If a boy grows up alone at the age of games and sports, and learns neither to play ball, nor row, nor sail, nor ride, nor skate, nor fish, nor shoot, probably he will be sedentary to the end of his days; and, though the best of opportunities be afforded him for learning these things later, it is a hundred to one but he will pass them by and shrink back from the effort of taking those necessary first steps the prospect of which, at an earlier age, would have filled him with eager delight. The sexual passion expires after a protracted reign; but it is well known that its peculiar manifestations in a given individual depend almost entirely on the habits he may form during the early period of its activity. Exposure to bad company then makes him a loose liver all his days; chastity kept at first makes the same easy later on. In all pedagogy the great thing is to strike the iron while hot, and to seize the wave of the pupil's interest in each successive subject before its ebb has come, so that knowledge may be got and a habit of skill acquired – a headway of interest, in short, secured, on which afterward the individual may float. There is a happy moment for fixing skill in drawing, for making boys collectors in natural history, and presently dissectors and botanists; then for initiating them into the harmonies of mechanics and the wonders of physical and chemical law. Later, introspective psychology and the metaphysical and religious mysteries take their turn; and, last of all, the drama of human affairs and worldly wisdom in the widest sense of the term. In each of us a saturation-point is soon reached in all these things; the impetus of our purely intellectual zeal expires, and unless the topic be one associated with some urgent personal need that keeps our wits constantly whetted about it, we settle into an equilibrium, and live on what we learned when our interest was fresh and instinctive, without adding to the store. Outside of their own business, the ideas gained by men before they are twenty-five are practically the only ideas they shall have in their lives. They *cannot* get anything new. Disinterested curiosity is past, the mental grooves and channels set, the power of assimilation gone. If by chance we ever do learn anything about some entirely new topic we are afflicted with a strange sense of insecurity, and we fear to advance a resolute opinion. But, with things learned in the plastic days of instinctive curiosity we never lose entirely our sense of being at home. There remains a kinship, a sentiment of intimate acquaintance, which, even when we know we have failed to keep abreast of the subject, flatters us with a sense of power over it, and makes us feel not altogether out of the pale.

Whatever individual exceptions might be cited to this are of the sort that "prove the rule."

To detect the moment of the instinctive readiness for the subject is, then, the first duty of every educator. As for the pupils, it would probably lead to a more earnest temper on the part of college students if they had less belief in their unlimited future intellectual potentialities, and could be brought to realize that whatever physics and political economy and philosophy they are now acquiring are, for better or worse, the physics and political economy and philosophy that will have to serve them to the end.

The natural conclusion to draw from this transiency of instincts is that most of them are implanted for the sake of giving rise to habits, and that, this purpose once accomplished, the instincts themselves, as such, have no *raison d'être* in the psychical economy, and consequently fade away. That occasionally an instinct should fade before circumstances permit of a habit being formed, or that, if the habit be formed, other factors than the pure instinct should modify its course, need not surprise us. Life is full of the imperfect adjustment to individual cases, of arrangements which, taking the species as a whole, are quite orderly and regular. Instinct cannot be expected to escape this general rule.

The most interesting thing possible now would be to test our principles by going through the human instincts in detail. But as I have already exceeded my allotted space, that must be reserved for another opportunity.

Source Endnotes

a "It would be very simple-minded to suppose that bees follow their queen, and protect her and care for her, because they are aware that without her the hive would become extinct. The odor or the aspect of their queen is manifestly agreeable to the bees – that is why they love her so. Does not all true love base itself on agreeable perceptions much more than on representations of utility?" Schneider: Thierische Wille, p. 187. *A priori,* there is no reason to suppose that *any* sensation might not in *some* animal cause *any* emotion and *any* impulse. To us it seems unnatural that an odor should directly excite anger or fear; or a color, lust. Yet there are creatures to which some smells are quite as frightful as any sounds, and very likely others to which color is as much a sexual irritant as form.

b In the instincts of mammals, and even of lower creatures, the uniformity and infallibility which, a generation ago, were considered as essential characters do not exist. The minuter study of recent years has found continuity, transition, variation, and mistake, wherever it has looked for them, and decided that what is called an instinct is usually only a tendency to act in a way of which the *average* is pretty constant, but which need not be mathematically "true." Cf. on this point, Darwin's Origin of Species; Romanes's Mental Evol., chaps. xi. to xvi., inch, and Appendix; Lindsay's Mind in Lower Animals, vol. i., 133–141 – ii., chaps. v., xx.; and Semper's Conditions of Existence in Animals, where a great many instances will be found.

c Spalding, *op. cit.,* p. 287.

d *Op. cit.,* p. 289.

e For the cases in full, see Mental Evolution in Animals, pp. 213–217.

Editorial Endnotes

1 *G. H. Schneider*: Georg Heinrich Schneider (1846–1904) was a German educator and philosopher who published monographs on the will in animals (*Der thierische Wille*, 1880) and humans (*Der menschliche Wille*, 1882).

2 "a priori *synthesis*": a concept developed in Immanuel Kant's *Critique of Pure Reason* (1781) in reference to knowledge obtained prior to, and independent of, experience and based on reason alone.

3 *Berkeley*: George Berkeley (1685–1753) was an Irish philosopher known for his philosophy of subjective idealism – developed in his *Treatise Concerning the Principles of Human Knowledge* (1710) – which held that the material world and everything in it only exist as mental phenomena.

4 *D. A. Spalding and Romanes*: these are British biologist and pioneering ethologist Douglas Alexander Spalding (1841–1877) and British physiologist George John Romanes (1848–1894). The latter's *Animal Intelligence* (1882) was an early example of comparative psychology and often referenced in debates about animal minds.

Editorial Headnote

8. Richard Lynch Garner, *The Speech of Monkeys*, New York: Charles L. Webster & Co., 1892.

To the amusement of much of the scientific establishment of the time, Richard Lynch Garner (1848–1920) was outspoken about his belief that the vocalisations of apes and monkeys could be made intelligible to humans and that the primates themselves could be taught to understand and use human languages – claims that were certain to raise some eyebrows in contemporary debates about the evolution and exceptionality of human language. In the 1880s, the eccentric autodidact primatologist spent much of his time in various American zoos to observe the habits of primates and produce phonographic recordings of their vocalisations, which he would play back to them to test their reactions. Fearing, however, that the study of captive animals might be insufficient, in 1892 Garner made his first of several trips to the French colony of Gabon to study primates in their natural habitat. His fieldwork in Africa caught the attention of the Smithsonian Institution and the New York Zoological Society – less for the results or prospects of his research than for his ability to procure much-sought-after chimpanzees, gorillas, and other "exotic" creatures for American zoos and museums (not averse to cultivating the image of the intrepid white hunter braving the dangers of the "Dark Continent," Garner's success fundamentally relied on local Gabonese labor and expertise).

In Gabon, Garner attempted to teach his young chimpanzee Moses and other captured or purchased apes to communicate with him. Though largely unsuccessful in his efforts at interspecies communication, in an 1896 book detailing his experiences in the French colony Garner reaffirms his belief in the linguistic abilities of nonhuman primates, insisting that ape vocalisations "have all the characteristics of true speech" because the ape speaker knew "the value of sound as a medium of thought" and was "conscious of the meaning of the sound used, and uses it with the definite purpose of conveying an idea to the one addressed."[1] In the final decade of his life, Garner appeared in a number of high-profile public performances in various parts of the U.S. with his chimpanzee Susie, with whom he claimed to be able to hold conversations. Susie never spoke, but this did little to diminish her entertainment value to the audiences flocking to Garner's presentations.[2]

Preceded by articles on the topic in respected magazines like *The Forum* and *North American Review*, Garner's first monograph on his studies of simians, *The Speech of Monkeys*, was published in the year of his first trip to Gabon. In the first half of the book, from which the following excerpts are taken, Garner recounts his interactions with a variety of (more or less cooperative) individual animals and his attempts, with the help of his phonograph, at deciphering the meanings of simian vocalisations.

Notes

1 Richard Lynch Garner, *Gorillas and Chimpanzees* (London: Osgood, McIlvaine & Co., 1896), 74.

2 Jeremy Rich, *Missing Links: The African and American Worlds of R. L. Garner, Primate Collector* (Athens: University of Georgia Press, 2012); Gregory Radick, *The Simian Tongue: The Long Debate about Animal Language* (Chicago: University of Chicago Press, 2007).

8

LEARNING THE SIMIAN TONGUE (1892)

Richard L. Garner

Source: Richard Lynch Garner, *The Speech of Monkeys* (New York: Charles L. Webster & Co., 1892), 3–21, 30–9, 57–67.

FROM childhood, I have believed that all kinds of animals have some mode of speech by which they can talk among their own kind; and I have often wondered why man has never tried to learn it. I have often wondered how it occurred to man to whistle to a horse or dog instead of using some sound more like their own, and even yet I am at a loss to know how such a sound has ever become a fixed means of calling these animals. I am not alone in my belief that all animals have some way to make known to others some certain things; but to my mind, the means have never been well defined.

About eight years ago, in the Cincinnati Zoological Garden, I was deeply impressed by the conduct of a number of monkeys occupying a cage with a huge, savage mandril, which they seemed very much to fear and dislike. By means of a wall the cage was divided into two compartments, through which was a small doorway, just large enough to allow the occupants of the cage to pass from one room to the other. The inner compartment of the cage was used for their winter quarters and sleeping-apartments; the outer, consisting simply of a well-constructed iron cage, was intended for exercise and summer occupancy. Every movement of this mandril seemed to be closely watched by the monkeys that were in a position to see him, and instantly reported to the others in the adjoining compartment. I watched them for hours, and felt assured that they had a form of speech by means of which they communicated with each other. During the time I remained, I discovered that a certain sound would invariably cause them to act in a certain way, and in the course of my visit I discovered that I could myself tell by the sounds the monkeys would make just what the mandril was doing – that is, I could tell whether he was asleep or whether he was moving about in his cage. Having interpreted one or two of these sounds, I felt inspired with the belief that I could learn them, and felt that the "key to the secret chamber" was within my grasp.

DOI: 10.4324/9781003032717-9

I regarded the task of learning the speech of monkeys as very much the same as learning that of some strange race of mankind – more difficult in the degree of its inferiority, but less in volume.

Year by year, as new ideas were revealed to me, new barriers arose, and I began to realize how great a task was mine. One difficulty was to utter the sounds I heard, another was to recall them, and yet another to translate them. But impelled by an inordinate hope and not discouraged by poor success, I continued my studies as best I could in the gardens of New York, Philadelphia, Cincinnati, and Chicago, and with such specimens as I could find from time to time with travelling shows, hand-organs, aboard some ship, or kept as a family pet. I must acknowledge my debt of gratitude to all these little creatures who have aided me in the study of their native tongue.

Having contended for some years with the difficulties mentioned, a new idea dawned upon me, and after maturely considering it, I felt assured of ultimate success. I went to Washington and proposed the novel experiment of acting as interpreter between two monkeys. Of course, this first evoked from the great fathers of science a smile of incredulity; but when I explained the means by which I expected to accomplish this, a shadow of seriousness came over the faces of those dignitaries to whom I first proposed the novel feat. I procured a phonograph[a] upon which to record the sounds of the monkeys. I separated two monkeys which had occupied the same cage together for some time, and placed them in separate rooms of the building, where they could not see or hear each other. I then arranged the phonograph near the cage of the female, and by various means induced her to utter a few sounds, which were recorded on the cylinder of the instrument. The machine was then placed near the cage containing the male and the record repeated to him, and his conduct closely studied. He gave evident signs of recognizing the sounds, and at once began a search for the mysterious monkey doing the talking. His perplexity at this strange affair cannot well be described. The familiar voice of his mate would induce him to approach, but that squeaking, chattering horn was a feature which he could not comprehend. He traced the sounds, however, to the source from which they came, and failing to find his mate, thrust his arm into the horn quite up to his shoulder, then withdrew it, and peeped into it again and again. The expressions of his face were indeed a study. I then secured a few sounds of his voice and delivered them to the female, who showed some signs of interest, but the record was very imperfect and her manner seemed quite indifferent. In this experiment for the first time in the history of language was the simian speech reduced to record, and while the results were not fully up to my hopes, they served to inspire me to further efforts to find the fountain-head from which flows out the great river of human speech. Having satisfied myself that each one recognized the sound made by the other when delivered through the phonograph, I felt rewarded for my labor and assured of the possibility of learning the language of monkeys. The faith of others was strengthened also, and while this experiment was very crude and imperfect, it served to convince me that my opinions were correct as to the speech of these animals.

In this case I noticed the defects which occurred in my work, and provided against them, as well as I could, for the future. Soon after this I went to Chicago and Cincinnati, where I made a variety of records of the sounds of a great number of monkeys, and among others I secured a splendid record of the two chimpanzees contained in the Cincinnati collection, which I brought home with me for study. The records that I made of various specimens of the simian race I repeated to myself over and over until I became familiar with them and learned to imitate a few of them, mostly by the use of mechanical devices. After having accomplished this, I returned to Chicago, and went at once to visit a small Capuchin monkey whose record had been my chief study. Standing near his cage I imitated a sound which I had translated "milk;" but from many tests I concluded it meant "food," which opinion has been somewhat modified by many later experiments that lead me to believe that he uses it in a still wider sense. It is difficult to find any formula of human speech equivalent to it. While the Capuchin uses it relating to food and sometimes to drink, I was unable to detect any difference in the sounds. He also seemed to connect the same sound to every kindly office done him and to use it as a kind of "Shibboleth." More recently, however, I have detected in the sound slight changes of inflection under different conditions, until I am now led to believe that the meaning of the word depends somewhat, if not wholly, on its modulation. The phonetic effect is rich and rather flute-like, and the word resembles somewhat the word "who." Its dominant is a pure vocal "u," sounded like "oo" in "too," which has a faint initial "wh," both elements of which are sounded, and the word ends with a vanishing "w." The literal formula by which I would represent it is "wh-oo-w." The word which I have translated "drink" begins with a faint guttural "ch," glides through a sound resembling the French diphthong "eu," and ends with a slight "y" sound, as in "ye."

So far I have found no trace of the English vowels "a," "i," or "o," unless it be in the sound emitted under stress of great alarm or in ease of assault, in which I find a close resemblance to the vowel "i" short, as in "it."

After having acquired a sound or two I extended my field of operations, and began to try my skill as a simian linguist on every specimen with which I came in contact.

In Charleston a gentleman owns a fine specimen of the brown Cebus[1] whose name is Jokes. He is naturally shy of strangers, but on my first visit to him I addressed him in his native tongue, and he really seemed to regard me very kindly; he would eat from my hand and allow me to caress him through the bars of his cage.

He eyed me with evident curiosity, but invariably responded to the word that I uttered in his own language. On my third visit to him I determined to try the effect of the peculiar sound of "alarm" or "assault" which I had learned from one of this species; but I cannot very well represent it in letters. While he was eating from my hand I gave this peculiar piercing note, and he instantly sprang to a perch in the top of his cage; thence in and out of his sleeping-apartment with great speed and almost wild with fear. As I repeated the sound his fears seemed to increase, until from a mere sense of compassion I desisted. No amount of coaxing would induce him to return to me or to accept any offer of peace which I could make. I retired to a distance of about twenty feet from his cage, and his master induced him to descend from the perch, which he did with the greatest reluctance and suspicion. I gave the sound again from where I stood, and it produced almost the same results as before. The monkey gave out a singular sound in response to my efforts to appease him, but refused to become reconciled.

After the lapse of eight or ten days I had not been able to reinstate myself in his good graces or to induce him to accept anything whatever from me. At this juncture I resorted to harsher means of bringing him to terms and began to threaten him with a rod. At first he resented this, but soon yielded and came down merely from fear. He would place the side of his head on the floor, put out his tongue, and utter a very plaintive sound having a slight interrogative inflection. At first this act quite defied interpretation; but during the same period I was visiting a little monkey called Jack. For strangers we were quite good friends, and he allowed me many liberties, which the family assured me he had uniformly refused to others.

On one of my visits he displayed his temper and made an attack upon me, because I refused to let go of a saucer from which I was feeding him some milk. I jerked him up by the chain and slapped him sharply, whereupon he instantly laid the side of his head on the floor, put out his tongue, and made just such a sound as Jokes had made a number of times before. It occurred to me that it was a sign of surrender, and many subsequent tests have confirmed this opinion.

Mrs. M. French Sheldon,[2] in her journey through East Africa, shot a small monkey in a forest near Lake Charla. She described to me how the little fellow stood high up in a tree and chattered to her in his sharp, musical voice until at the crack of her gun he fell mortally wounded. When he was laid dying at her feet, he turned his bright little eyes pleadingly upon her as if to ask for pity. Touched by his appeal, she took the little creature in her arms to try to soothe him. Again and again he would touch his tongue to her hand as if kissing it, and seemed to wish in the hour of death to be caressed even by the hand that slew him, and which had taken from him without reward that life which could be of no value except to spare to the wild forest where his kindred monkeys live.

This peculiar mode of expressing submission seems to be very widely used, and from her description of the actions of that monkey his conduct must have been identical with that of the Cebus, and to my mind may justly be interpreted to mean, "Pity me; I will not harm you." I have recently learned that a Scotch naturalist, commenting on my description of this act and its meaning, quite agrees with me, and states that he has observed the same thing in other species of monkeys.

DURING a period of many weeks I visited Jokes almost daily, but after the lapse of more than two months I had not won him back nor quieted his suspicions against me. On my approach he would manifest great fear and go through the act of humiliation described above. I observed that he entertained an intense hatred for a negro boy on the place, who teased and vexed him on all occasions. I had the boy come near the cage, and Jokes fairly raved with anger. I took a stick and pretended to beat the boy, and this delighted Jokes very greatly. I held the boy near enough to the cage to allow the monkey to scratch and pull his clothes, and this would fill his little simian soul with joy. I would then release the boy, and to the evident pleasure of Jokes would drive him away by throwing wads of paper at him. I repeated this a number of times, and by such means we again became the best of friends. After each encounter with the boy Jokes would come to the bars, touch my hand with his tongue, chatter and play with my fingers, and show every sign of confidence and friendship. He always warned me of the approach of any one, and his conduct toward them was largely governed by my own. He never failed after this to salute me with the sound described in the first chapter.

About the same time I paid a few visits to another little monkey of the same species, named Jennie. Her master had warned me in advance that she was not well disposed toward strangers. At my request he had her chained in a small side yard, which he forbade any of the family entering. When I approached the little lady for the first time I gave her the usual salutation, which she responded to and seemed to understand. I unceremoniously sat down by her side and fed her from my hands. She eyed me with evident interest and curiosity, while I studied her every act and expression. During the process of this mutual investigation, a negro girl who lived with the family, overcome by curiosity, stealthily entered the yard and came up within a few feet of us. I determined to sacrifice this girl upon the altar of science; so I arose and placed her between the monkey and myself and vigorously sounded the alarm or menace. Jennie flew into a fury, while I continued to sound the alarm and at the same time pretended to attack the girl with a club and some paper wads, thus causing the monkey to believe that the girl had uttered the alarm and made the assault. I then drove the girl from the yard with a great show of violence, and for days afterward she could not feed or approach the little simian. This confirmed my opinion of the meaning of the sound, which can be fairly imitated by placing the back of the hand gently on the mouth and kissing it with great force, prolonging the sound for some seconds. This imitation, however, is indifferent, and its quality is especially noticeable when analyzed on the phonograph. The pitch corresponds to the highest "F" sharp on the piano, while the word "drink" is two octaves lower and the word "food" three.

On one occasion I visited the garden in Cincinnati, and found in a cage a small Capuchin, to whom I gave the name Banquo. It was near night and the visitors had left the house, and the little monkey, worried out by the day's annoyance from visitors, sat quietly in the back of his cage, as though he was glad another day was done. I approached the cage and uttered the sound which I have described and translated "drink." My first effort caught his attention and caused him to turn and look at me. He then arose and answered me with the same word, and came at once to the front of the cage. He looked at me as if in doubt, and I repeated the word. He responded with the same and turned to a small pan in his cage, which he took up and placed near the door through which the keeper usually passed his food, returned to me, and uttered the word again. I asked the keeper for some milk, which he did not have, but brought me some water instead. The efforts of my little simian friend to secure the glass were very earnest, and his pleading manner and tone assured me of his extreme thirst I allowed him to dip his hand into the glass, and he would then lick the water from his fingers and reach again, I kept the glass out of reach of his hand, and he would repeat the sound earnestly and look at me beseechingly as if to say: "Please give me some more." I was thus convinced that the word which I had translated "milk" must also mean "water," and from this and other tests I at last determined that it meant "drink" in its broad sense, and possibly "thirst." It evidently expressed his desire for something with which to allay his thirst. The sound is very difficult to imitate and quite impossible to write exactly.

On one of my visits to the Chicago garden I stood with my side to a cage containing a small Capuchin, and gave the sound which I have translated "milk." It caused him to turn and look at me, and on repeating the sound a few times he answered me very distinctly with the same, picking up the pan from which he usually drank, and as I repeated the word he brought the pan to the front of the cage, set it down and came up to the bars, and uttered the word distinctly. I had not shown him any milk or any kind of food; but the man in charge, at my request, brought me some milk, which I gave to him. He drank it with great delight; then looked at me and held up his pan, repeating the sound. I am quite sure that he used the same sound each time that he wanted milk. During this same visit, I tried many experiments with the word which I am now convinced means "food" or "hunger." And I was led to the belief that he used the same word for apple, carrot, bread, and banana; but a few later experiments have led me to modify this view in a measure, since the phonograph shows me slight variations of the sound, and I now think it probable that these faint inflections may possibly indicate a difference in the kinds of food he has in mind. However, they usually recognize this sound, even when poorly imitated. I am impressed with the firm belief that in this word I have found the clew to the great secret of speech; and while I have taken only one short step in the direction of its solution, I have pointed out the way which leads to it.

In the fall of 1891 I visited New York for the purpose of experimenting with the monkeys in Central Park.[3] Early one morning I repaired to the monkey-house, and for the first time approached a cage containing five brown Capuchins, whom I saluted with the word which I have translated "food," and which seems to be an "open-sesame" to the hearts of all monkeys of this species. On delivering this word one of them responded promptly and came to the front of the cage. I repeated it two or three times and the remaining four came to the front, and as I thrust my fingers through the bars of the cage they took hold of them and began playing with great familiarity and apparent pleasure. They seemed to recognize the sound and to realize that it had been delivered to them by myself. Whether they regarded me as a great ape, monkey, or some other kind of animal speaking their tongue, I do not know. But they evidently understood the sound, though up to this time I had shown them no food or water. A little later I secured some apples and carrots and gave them in small bits in response to their continual requests for food, and this further confirmed my belief that I had translated the word correctly.

This was gratifying to me in view of the fact that I was accompanied by two gentlemen who had been permitted to witness the experiment, and it was evident to them that the monkeys understood the sound. I placed the phonograph in order and made a record of the sound, which I preserved for study. After an absence of some days, I returned to the park and went to the monkey-house. They recognized me as I entered the door, notwithstanding there were many visitors present. They began begging me to come to their cage, which I did, and gave them my hand to play with. One of them in particular, whose name is McGinty, showed every sign of pleasure at my visit; he would play with my fingers, hug them, and caress them in the most affectionate manner. Another occupant of the same cage had shown a disposition to become friendly with me, and on this occasion came bravely to the bars of the cage and showed a desire to share the pleasure of my visit with his little simian brother. But this was denied him on any terms by McGinty, who pounced upon him and drove him away, as he also did the other monkeys in the cage in order to monopolize my entire society himself. He refused to allow any other inmate of the cage to receive my caresses or any part of the food that I had brought them.

I spent the past winter in Washington and New York, much of the time in company with these little creatures, and have made many novel and curious experiments, some of which have resulted in surprises to myself. Among the facts which I have obtained, I may state that certain monkeys can count three; that they discern values by quantity and by number; that they have favorite colors and are pleased with some musical sounds.

In the Washington collection there is a Capuchin by the name of Pedro. When I first visited this bright little monkey he occupied a cage in common with several other monkeys of different kinds. All of them seemed to impose upon little Pedro, and a mischievous young spider-monkey found special delight in catching him by the tail and dragging him around the floor of the cage. I interfered on behalf of Pedro and drove the spider-monkey away. On account of this Pedro soon began to look upon me as his benefactor, and when he would see me he would scream and beg for me to come to him. I induced the keeper to place him in a small cage by himself, and this he seemed to appreciate very much. When I would go to record his sounds on the phonograph, I would hold him in one hand while he would take the tube in his tiny black hands, hold it close up to his mouth, and talk into it just like a good little boy who knew what to do and how to do it. He would sometimes laugh, and always chatter to me as long as he could see me. He would sit on my hand and kiss my cheeks, put his mouth up to my ear and chatter just as though he knew what my ears were for. He was quite fond of the head-keeper and also of the director; but he entertained a great dislike for one of the assistant keepers, and he has often told me some very bad things about that man, though I could not understand what he said. I shall long remember how this dear little monkey would cuddle up under my chin and try so hard to make me understand some sad story which seemed to be the burden of his life. He readily understood the sounds of his own speech which I repeated to him, and I have made some of the best records of his voice that I have ever succeeded in making of any monkey, some of which I have preserved up to this time. They present a wide range of sounds, and I have studied them with special care and pleasure, because I knew that they were addressed to me in person; and being aware that the little creature was uttering these sounds to me with the hope that I would understand them, I was more anxious to learn just what he really said to me in this record than if it had contained only some casual remark not addressed to me. This little simian was born in the Amazon Valley, in Brazil, and was named for the late emperor.

A short time ago I borrowed from a dealer in Washington a little Capuchin called Puck, and had him sent to my apartments, where I kept a phonograph. I placed the cage in front of the machine upon which I had adjusted the horn and had placed the record of my little friend Pedro. I concealed myself in an adjoining room, where I could watch the conduct of my subject through a small hole in the door. I had a string attached to the lever of the machine and drawn taut through another hole in the door, so that I could start the machine at any desired moment, and at the same time avoid attracting the attention of the monkey, either by my presence or by allowing him to see anything move. After a time, when everything was quiet, I set the machine in motion, and treated him to a phonographic recital by little Pedro. This speech was distinctly delivered through the horn to Puck, from whose actions it was evident that he recognized it as the voice of one of his tribe. He looked at the horn in surprise and made a sound or two, glanced around the room, and again uttered a couple of sounds as he retired from the horn apparently somewhat afraid. Again the horn delivered some exclamations in a pure Capuchin dialect, which Puck seemed to regard as sounds of some importance. He cautiously advanced and made a feeble response, but a quick, sharp sound from the horn seemed to startle him; and failing to find any trace of a monkey except the sound of a voice, he looked at the horn with evident suspicion, and scarcely ventured to answer any sound it made. When I had delivered to him the contents of the record I entered the room again, and this seemed to afford him some relief.

A little later I adjusted my apparatus for another trial, and this time I hung a small mirror just above the mouth of the horn. Then retiring again from the room I left him to examine his new surroundings, and he soon discovered the new monkey in the glass and began to caress it and chatter to it. After a while I started the phonograph again by means of the string, and when the horn began to deliver its simian oration, it appeared to disconcert and perplex Puck. He would look at the image in the glass, then he would look into the horn; he would retire with a feeble grunt and a kind of inquisitive grin, showing his little white teeth, and acting as though in doubt whether to regard the affair as a joke or to treat it as a grim and scientific fact. His voice and actions were exactly like those of a child declaring in words that he was not afraid, but betraying fear in every act, and finally blending his feelings into a genuine cry. Puck did not cry, but the evidence of fear made the grin on his face rather ghostly. Again he would approach the mirror, then listen to the sounds which came from the horn, and it appeared from his conduct that there was a conflict somewhere. It was evident that he did not believe that the monkey which he saw in the glass was making the sounds which came from the horn. He repeatedly put his mouth to the glass and caressed the image which he saw there, and at the same time showed a grave suspicion and some concern about the one which he heard in the horn, and tried to keep away from it as much as possible. His conduct in this case was a source of surprise to me, as the sounds contained in the record which I had repeated to him were all uttered in a mood of anxious, earnest entreaty, which to me seemed to contain no sound of anger, warning, or alarm, but which, on the contrary, I had interpreted as a kind of love-speech, full of music and tenderness. I had not learned the exact meaning of any one of the sounds contained in this cylinder, but had ascribed in a collective and general way such a meaning to this speech. But from Puck's conduct I was led to believe that it was a general complaint of some kind against those monkeys in that other cage who had made life a burden to little Pedro. One thing was clear to my mind, and that is that Puck interpreted the actions of the monkey which he saw in the glass to mean one thing, and the sounds which he heard from the horn to mean quite another.

I do not think that their language is capable of shading sentences into narrative or giving any detail in a complaint, for I have never seen anything yet among them that would justify one in ascribing to them so high a type of speech; but in terms of general grievance it may have conveyed to Puck the idea of a monkey in distress, and hence his desire to avoid it; while the image in the glass presented to him a picture of his own mood, and he therefore had no cause to shun it. I do think, however, that the present form of speech used by monkeys is developed far above a mere series of grunts and groans, and that some species among them have much more copious and expressive forms of speech than others do. From many experiments with the phonograph, I am prepared to say with certainty that some have much higher phonetic types than others do. I have traced some slight inflections which I think beyond a doubt modify the values of their sounds. I find monkeys who do not make certain inflections at all, although the phonation of a species is generally uniform in other respects. In some cases it seems to me that the inflections differ slightly in the same species, but long and constant association seems to unify these dialects in some degree, very much the same as like causes blend and unify the dialects of human speech. I have found one instance in which a Capuchin had acquired two sounds which strictly belonged to the tongue of the white-faced Cebus. I was surprised when I heard him utter the sounds, and thought at first that these sounds were common to the speech of both varieties, but on inquiry I found that he had been confined in a cage with the white-face for nearly four years, and hence my belief that he acquired them during that time.

The most remarkable case which has come under my observation is one in which a young white-face has acquired the sound which means food in the Capuchin tongue. This event occurred under my own eyes. I regard this case as so noteworthy, being attended by such conditions as to show that the monkey had a motive in learning the sound, that I shall relate it in detail.

In the room where the monkeys were kept by a dealer in Washington, there was a cage which contained a young white-faced Cebus of rather more than average intelligence. He was a quiet, sedate, and thoughtful little monkey, whose gray hair and beard gave him quite a venerable aspect, and for this reason I called him Darwin. From some cause unknown to me he was afraid of me, and I showed him but little attention. On the same shelf and in an adjacent cage lived the little Capuchin, Puck. The cages were only separated by an open wire partition through which they could easily see and hear each other. For some weeks I visited Puck almost daily, and in response to his sound for food I always supplied him with some nuts, banana, or other food. I never gave him any of these things to eat unless he would ask me for them in his own speech. On one of my visits my attention was attracted by little Darwin, who was uttering a strange sound which I had never before heard one of his species utter. I did not recognize the sound at first, but very soon discovered that it was intended to imitate the sound of the Capuchin, in response to which I always gave Puck some nice morsel of food. Darwin had undoubtedly observed that this sound made by Puck was always rewarded with something good to eat, and his evident motive was to secure a like reward. After this I always gave him some food in acknowledgment of his efforts, and I observed, from day to day, that he improved in making this sound until at last it could scarcely be detected from the sound made by Puck. This was accomplished within a period of less than six weeks from my first visit. In this case, at least, I have seen one step taken by a monkey in learning the tongue of another. This was most interesting to me in view of the fact that I had long believed, and had announced as my belief that no monkey ever acquired the sounds made by another species, or, indeed, ever tried to do so. I admit, however, that this one instance alone is sufficient to cause me to recede from a conclusion thus rendered untenable, and the short time in which this one feat was accomplished would indicate that the difficulty was not so great as I had regarded it. I still regard it as a rule, however, that monkeys do not learn each other's speech, but the rule is not without exceptions.

I have observed and called attention to the fact, that when two monkeys of different species are caged together, each one will learn to understand the speech of the other, but as a rule will not try to speak it. When he replies at all, it is always in his own vernacular. I wish to impress the fact that monkeys do not generally carry on a connected conversation. Their speech is usually limited to a single sound or remark, which is replied to in the same manner, and to suppose that their conversations are elaborate or of a highly social character is to go beyond the bounds of reason. This is the respect in which the masses fail to understand the real nature of the speech of monkeys or other animals.

In my intercourse with these little creatures, I cannot forget how often I have caught the spirit of their tones when no ray of meaning as mere words of speech had dawned upon me, and it is partly through such means that I have been able to interpret them. As a rule each act of a monkey is attended by some sound, and each sound by some act, which to another monkey of the same species always means a certain thing. There are many cases perhaps, in which acquired words or shades of dialect are not quite clear to them, just as we often find in human speech; but monkeys appear to meet this difficulty and overcome it just as men do. They talk with one another on a limited number of subjects, but in very few words, which they frequently repeat if necessary. Their language is purely one of sounds, and while those sounds are accompanied by signs, generally, I think, they are quite able to get along better with the sounds alone than with the signs alone. The rules by which we may interpret the sounds of simian speech are the same as those by which we would interpret human speech. If you should be cast away upon an island inhabited by some strange race of people whose speech was so unlike your own that you could not understand a single word of it, you would watch the actions of those people, and see what act they did in connection with any sound they made; and in this way you would gradually learn to associate a certain sound with a certain act, until at last you would be able to understand the sound without seeing the act at all; and such is the simple line I have pursued in the study of the speech of this little race – only I have been compelled to resort to some very novel means of doing my part of the talking. Since I have been so long associated with them, I have learned to know in many cases what act they will perform in response to certain sounds; and as I grow more and more familiar with these sounds, I become better able to distinguish them, just as we do with human speech.

Until recently I have believed that their sounds were so limited in number as to preclude any specific terms in their vocabulary; but now I am inclined to modify this opinion somewhat, as I have reason to believe that they have some specific terms, such as a word for "monkey," another word for "fruit," and so on. They do not specify, perhaps, the various kinds of monkeys, but monkeys in general, in contradistinction to birds or dogs. Their word for fruit does not specify the kind, but only means fruit in a collective sense, and only as a kind of food. I am not positive as yet that their specific terms may even go so far as this; but I infer that such may be the case from one fact which I have observed in my experiments. When I show a monkey his image in a mirror, he utters a sound on seeing it – especially if he has been kept away from other monkeys for a long time; and all monkeys of the same species, so far as I have observed, under like conditions use the same sound and address it in the same way to the image in the glass. In a few instances I have seen strange monkeys brought in contact with each other, and have observed that they use this same sound on their first meeting. The sound is always uttered in a low, soft tone, and appears to have the value of a salutation. When kept in a cage with other monkeys they do not appear to salute the image in the glass, but chatter to it, and show less surprise at seeing it than in cases where they have been kept alone for some time.

[. . .]

One of the most certain of my discoveries in the simian speech is the negative sign and the word "no." The sign is made by shaking the head from side to side in a fashion almost exactly like that used by man to express the same idea. I have no longer any doubt of the intent and meaning of this sign, and the many tests to which I have subjected it compel me to accept the result as final.

A little more than a year ago my attention was called to this sign by the children who own the little Capuchin, Jack, in Charleston. A number of times they said to him, in my presence: "Jack, you must go to bed." At which he would shake his little black head, as if he really did not wish to comply. I watched this with great interest; but it was my belief at that time that he had been trained to do this, and that the sign did not really signify to him anything at all. The children, however, declared to me that he really meant "no." To believe that he meant this would presuppose that he understood the combination of words quoted, and this was beyond the limits of my faith, although it was certain that a repetition of the sentence always elicited from him the same sign, which indicated that he recognized it as the same sentence or combination of sounds, and gave it the same reply each time. I concluded that he had been taught to associate this sign with some sound – for instance, "bed" or "go;" but since that time I have found the sign to be almost universal with this species of monkey, and they use the sign to express negation. I have seen them use the sign in response to certain things which were wholly new to them, but where the idea was clear to them and they desired to express dissent. The fact that this sign is common to both man and simian I regard as more than a mere coincidence, and I believe that in this sign I have found the psycho-physical basis of expression.

I have made scores of experiments on this subject, and I find this sign a fixed factor of expression. In one case where I tried to induce a monkey to allow me to take him into my hands from the hand of his master, he would shake his head each time, and make a peculiar sound somewhat like a suppressed cluck. I would try to coax him with nuts, in response to which he would make the same sound and sign each time, and his actions showed beyond all controversy his intention. I had taught a monkey to drink milk from a bottle by sucking it through a rubber nipple, and after he had satisfied his thirst, when I would try to force the bottle to his lips he would invariably respond by a shake of the head in the manner described, and at the same time utter a clucking sound. I tried many similar experiments with three or four other monkeys, and secured the same result in each case. In another instance, where a monkey was confined in a small cage, so that I could easily catch him in order to tame him by handling, when I would put my hand into the cage to catch him he would shake his head in this manner and accompany the act by a plaintive sound, which was so touching that I could not obtain my own consent to persecute the little prisoner by compelling him to submit to my caresses. I have found that the little rogue McGinty, in Central Park, does the same thing at times when I go into the cage and attempt to put my hands on him, and especially when he has taken refuge in a corner to nurse his jealousy. While I remain outside the cage he is so devoted to me that he will scarcely leave me to get something to eat; but when I enter the cage and reach out my hand toward him, he will shake his little head and utter that peculiar clucking sound. Many of these tests I have repeated over and over with the same results, and noting the conditions at the time, I am thoroughly convinced that the sign and sound mean "no." I have observed that this sigh is always made in the same manner; but sometimes it is accompanied by a clucking sound, while at other times it is a soft, whimpering sound, almost like a low, plaintive whistle, The sign is frequently used without the sound at all; and I must impress it upon my reader that these results do not always present themselves in every experiment, as much depends upon the mood and surroundings of the subject, I have found that one advantage is to have the monkey confined in a very small cage, as otherwise he will turn away and get out of your reach when you press anything upon him that he does not want, I have also found much better results by having the monkey alone, and where he can neither see nor hear other monkeys.

Having discovered the sign of negation among the simians, I began an investigation to ascertain how far it could be found among the races of mankind. I have carried my search far beyond the limits of local inquiry, and up to this time I have found only a few trifling exceptions in the use of this sign among all the races of men, and those few exceptions are found among the Caucasian race, and appear to be confined to Southern Europe. I have heard that among certain island tribes of Polynesia these signs are reversed; but I have been assured by two officers of the English navy and two of the United States navy, who have visited the islands in question, that such is not the case. Among the Indians, Mongolians, and Negroes I have found no noteworthy exceptions. I have inquired among mothers who have raised families to ascertain where they first observed this sign as an expression among their children; and from the consensus of opinion it appears that this is about the first sign used by infants to express negation.

I have not found the positive sign, or sign of affirmation by a nod of the head, to be so general; yet it has a wide range within the human family, and appears to be used to some extent among the lower primates.

Seeking a source from which these signs may have originated, I have concluded that they may arise from two circumstances: The negative sign, doubtless, comes from an effort to turn the head away from something which is not desired, and that, with such an intent, it has gradually crystallized into an instinctive expression of negation or refusal; while the nod of affirmation or approval may have grown out of the intuitive lowering of the head as an act of submission or acquiescence, or from reaching the head forward to receive something desired, or it may have come from these two causes conjointly.

This is only one of a great many points in which the speech of simians coincides with that of man. It is true we have no letters in our alphabet with which to represent the sounds of their speech, nor have we the phonetic equivalence of their speech in our language; but it is also true that our alphabet does not fully represent or correctly express the entire phonetic range of our own speech; but the fact that our speech is not founded upon the same phonetic basis, or built up into the same phonetic structures, is no reason that their speech is not as truly speech as our own. That there are no letters in any alphabet which represent the phonetic elements of simian speech is doubtless due to the fact that there has never been any demand for such; but the same genius that invented an alphabet for human speech, actuated by the same motives and led by the same incentives, could as easily invent an alphabet for simian speech. It is not only true that the phonetic elements of our language are not represented by the characters of our alphabet, but the same is true to some extent of our words which do not quite keep pace with human thought. In the higher types of human speech there are thousands of words and ideas which cannot be translated into or expressed by any savage tongue, because no savage ever had use for them and no savage tongue contains their equivalence. The growth of speech is always measured by the growth of mind. They are not always of the same extent, but always bear a common ratio. It is a mental product, and must be equal to the task of coining thoughts into words. It is essential to all social order, and no community could long survive as such without it. It is as much the product of mind and matter as salt is the product of chlorine and sodium.

Source Endnote

a While I use the word phonograph throughout this work, I must state that many of my experiments were made with the graphophone.

Editorial Endnotes

1 *Cebus*: a genus that comprises several kinds of capuchin monkeys.

2 *Mrs. M. French Sheldon*: Mary French Sheldon (1847–1936), an American traveler, explorer, and author. Her experiences in the Belgian Congo and other parts of Africa are recounted in *Sultan to Sultan* (1892).

3 *the monkeys in Central Park*: this refers to New York's Central Park Menagerie (later Central Park Zoo; see J. W. Smith, Vol. 5).

Editorial Headnote

9. Edward Lee Thorndike, "Do Animals Reason?" (1899)

9.1 Edward Lee Thorndike, "Do Animals Reason?" *Popular Science Monthly* 55 (August 1899): 480–490.

9.2 "Correspondence: 'Do Animals Reason?'" *Popular Science Monthly* 55 (October 1899): 843–847.

With the founding of *Popular Science Monthly*, New York science writer and educator Edward Livingston Youmans finally realised his idea of a publication that would bring scientific knowledge and news to an audience of interested laypeople. Serving as the magazine's editor until his death in 1887, Youmans had previously worked as scientific advisor for the New York publishing house Appleton, which would also publish Youmans' new magazine. The May 1872 inaugural issue of *Popular Science Monthly* opened with the first of a 14-part series on "The Study of Sociology" by Herbert Spencer, then already a popular figure in the United States, and Spencer's work would also feature heavily in future issues, no doubt contributing significantly to the magazine's success. With its transatlantic scope, the magazine regularly introduced its readers to the work of European authors in the form of article reprints or translations.[1]

Animal minds and the forms and degrees of intelligence exhibited by different types of animals were a recurring topic in the pages of the *Popular Science Monthly*. Columbia University psychologist Edward Lee Thorndike (1874–1949), a former student of William James's at Harvard, approached this topic from a perspective that departed significantly from earlier texts. While the question "Do Animals Reason?" was, of course, very much a conventional one in the final decade of the nineteenth century, the way in which Thorndike sought to answer it – through laboratory science – was not. Thorndike considered authors who ascribed reasoning abilities to animals as part of a tradition of misleading anthropomorphism, a tradition he intended to invalidate by systematically observing animal behavior in pre-prepared and manipulable experimental settings.

For this purpose, Thorndike invented (apparently in James's basement) a contraption he called the "problem" or "puzzle box" – the precursor of the "Skinner box" later used by famous behaviorist Burrhus Frederic Skinner in his own experiments on animal learning with rats and pigeons. Thorndike's *Popular Science* article describes the arrangement and results of his experiments, in which (hungry) cats or dogs would find themselves trapped in one of his puzzle boxes but could free themselves by using a button or lever or similar mechanism, with food placed outside the box to provide additional motivation. Thorndike would then observe how the animals behaved, how quickly they managed to escape from the box, and if they learned from their experience and were able to free themselves more quickly when placed in the same situation again. Judging from the readers' responses, however, not everyone was convinced by Thorndike's experiment and the conclusions about animal intelligence he drew from it.[2]

Notes

1 See Frank Luther Mott, *A History of American Magazines*, vol. 3: 1865–1885 (Cambridge: Harvard University Press, 1938), 495–99.

2 Geraldine Jonçich Clifford, "Thorndike, Edward Lee," in *American National Biography*, ed. John A. Garraty and Mark C. Carnes, vol. 21 (New York: Oxford University Press, 1999), 604–5. On Thorndike's invention, see John C. Burnham, "Thorndike's Puzzle Boxes," *Journal of the History of the Behavioral Sciences* 8, no. 2 (1972): 159–67.

9

DO ANIMALS REASON? (1899)

Edward L. Thorndike (and correspondence)

Sources:

9.1 Edward Lee Thorndike, "Do Animals Reason?" *Popular Science Monthly* 55 (August 1899): 480–490.
9.2 "Correspondence: 'Do Animals Reason?'" *Popular Science Monthly* 55 (October 1899): 843–847.

9.1 DO ANIMALS REASON?

PROBABLY every reader who owns a dog or cat has already answered the question which forms our title, and the chance is ten to one that he has answered, "Yes." In spite of the declarations of the psychologists from Descartes to Lloyd Morgan,[1] the man who likes his dog and the woman who pets a cat persist in the belief that their pets carry on thinking processes similar, at least in kind, to our own. And if one has nothing more to say for the opposite view than the stock arguments of the psychologists, he will make few converts. A series of experiments carried on for two years have, I hope, given me some things more to say – some things which may interest the believer in reason in animals, even if they do not convert him.

In trying to find out what sort of thinking animals were capable of I adopted a novel but very simple method. Dogs and cats were shut up, when hungry, in inclosures from which they could escape by performing some simple act, such as pulling a wire loop, stepping on a platform or lever, clawing down a string stretched across the inclosure, turning a wooden button, etc. In each case the act set in play some simple mechanism which opened the door. A piece of fish or meat outside the inclosure furnished the motive for their attempts to escape. The inclosures for the cats were wooden boxes, in shape and appearance like the one pictured in Figure 9.1, and were about 20 × 15 × 12 inches in size. The boxes for the dogs (who were rather small, weighing on the average about thirty pounds) were about 40 × 22 × 22. By means of such experiments we put animals in situations seeming almost sure to call forth any reasoning powers they possess. On the days when the experiments were taking place they were practically utterly hungry, and so had the best reasons for making every effort to escape. As a fact, their conduct when shut up in these boxes showed the utmost eagerness to get out and get at the much-needed food. Moreover, the actions required and the thinking involved are such as the stories told about intelligent animals credit them with, and, on the other hand, are not far removed from the acts and feelings required in the ordinary course of

DOI: 10.4324/9781003032717-10

animal life. It would be foolish to deny reason to an animal because he failed to do something (e. g., a mathematical computation) which in the nature of his life he would never be likely to think about, or which his bones and muscles were not fitted to perform, or which, even by those who credit him with reason, he is never supposed to do. So the experiments were arranged with a view of giving reasoning every chance to display itself if it existed.

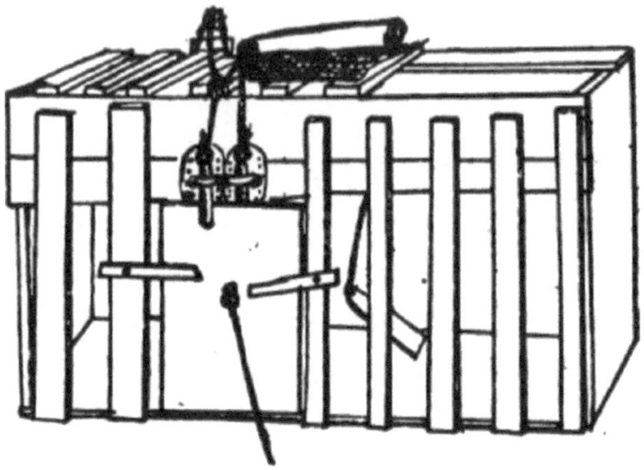

Figure 9.1

What, now, would we expect to observe if a *reasoning* animal, who is surely eager to get out, is put, for example, into a box with a door arranged so as to fall open when a wooden button holding it at the top (on the inside) is turned from its vertical to a horizontal position? We should expect that he would first try to claw the whole box apart or to crawl out between the bars. He would soon realize the futility of this and stop to consider. He might then think of the button as being the vital point, or of having seen doors open when buttons were turned. He might then poke or claw it around. If after he had eaten the bit of fish outside he was immediately put in the box again he ought to remember what he had done before, and at once attack the button, and so ever after. It might very well be that he would not, when in the box for the first time, be able to reason out the way to escape. But suppose that, in clawing, biting, trying to crawl through holes, etc., he happened to turn the button and so escape. He ought, then, if at once put in again, this time to perform deliberately the act which he had in the first trial hit upon accidentally. This one would expect to see if the animal *did* reason. What do we really see?

To save time we may confine ourselves to a description of the twelve cats experimented with, adding now that the dogs presented no difference in behavior which would modify our conclusions. The behavior of all but No. 11 and No. 13 was practically the same. When put into the box the cat would show evident signs of discomfort and of an impulse to escape from confinement. It tries to squeeze through any opening; it claws and bites at the bars; it thrusts its paws out through any opening, and claws at everything it reaches; it continues its efforts when it strikes anything loose and shaky; it may claw at things in the box. The vigor with which it struggles is extraordinary. For eight or ten minutes it will claw and bite and squeeze incessantly. With No. 13, an old cat, and No. 11, an uncommonly sluggish cat, the behavior was different. They did not struggle vigorously or continually. (In the experiments it was found that these two would stay quietly in the box for hours, and I therefore let them out myself a few times, so that they might associate the fact of being outside with the fact of eating, and so desire to escape. When this was done, they tried to get out like the rest.) In all cases the instinctive struggle is likely to succeed in leading the cat accidentally to turn the button and so escape, for the cat claws and bites all over the box. These general clawings, bitings, and squeezings are of course instinctive, not premeditated. The cats will do the same if in a box with absolutely no chance for escape, or in a basket without even an opening – will do them, that is, when they are the foolishest things to do. The cats do these acts for just the same reason that they suck when young, propagate when older, or eat meat when they smell it.

Each of the twelve cats was tried in a number of different boxes, and in no case did I see anything that even looked like thoughtful contemplation of the situation or deliberation over possible ways of winning freedom. Furthermore, in every case any cat who had thus accidentally hit upon the proper act was, after he had eaten the bit of fish outside, immediately put back into the box. Did he then think of how he had got out before, and at once or after a time of thinking repeat the act? By no means. He bursts out into the same instinctive activities as before, and may even fail this time to get out at all, or until a much *longer* period of miscellaneous scrabbling at last happens to include the particular clawing or-poking which works the mechanism. If one repeats the process, keeps putting the cat back into the box after each success, the amount of the useless action gradually decreases, the right movement is made sooner and sooner, until finally it is done as soon as the cat is put in.

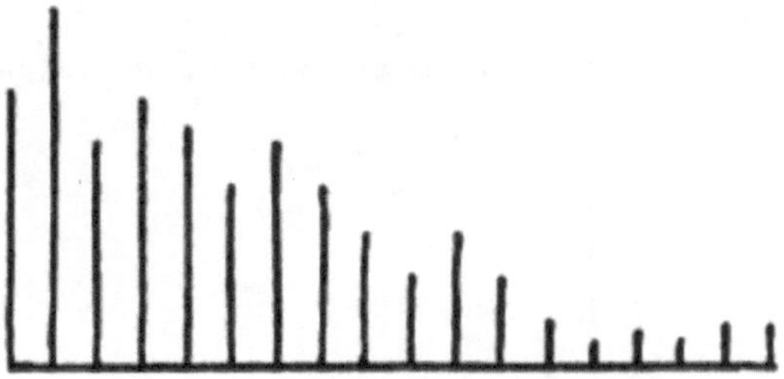

Figure 9.2

This sort of a history is not the history of a reasoning animal. It is the history of an animal who meets a certain situation with a lot of instinctive acts. Included without design among these acts is one which brings freedom and food. The pleasurable result of this one gradually stamps it in in connection with the situation "confinement in that box," while their failure to result in any pleasure gradually stamps out all the useless bitings, clawings, and squeezings. Thus, little by little, the one act becomes more and more likely to be done in that situation, while the others slowly vanish. This history represents the wearing smooth of a path in the brain, not the decisions of a rational consciousness.

We can express graphically the difference between the conduct of a reasoning animal and that of these dogs and cats by means of a time-curve. If, for instance, we let perpendiculars to a horizontal line represent each one trial in the box, and let their heights represent in each trial the time it took the animal to escape (each three millimetres equaling ten seconds), the accompanying figure (Figure 9.2) will tell the story of a cat which, when first put in, took sixty seconds to get out; in the second trial, eighty; in the third, fifty; in the fourth, sixty; in the fifth, fifty; in the sixth, forty, etc. This figure represents what did actually happen with one cat in learning a very easy act. Suppose the cat had, after the third accidental success, been able to reason. She would then have the next time and in all succeeding times performed the act as soon as put in, and the figure would have been such as we see in Figure 9.3. The thing is still clearer if, instead of drawing in the perpendiculars, we draw only a line joining their tops. Figure 9.4 shows, then, the curve for the real history, and Figure 9.5 shows the abrupt descent, due to a rational comprehension of the situation. I kept an accurate record of the time, in seconds, taken in every trial by every cat in every box, and in them all there appears no evidence for the presence of even the little reasoning that "what let me out of this box three seconds ago will let me out now." Surely, if an animal could reason he would, after ten or eleven accidental successes, think what he had been doing, and at the eleventh or twelfth trial would at once perform the act. But no! The slope of the curves, as one may see in the specimens shown in Figure 9.6, is always gradual. So, in saying that the behavior of the animals throughout the experiments gave no sign of the presence of reasoning I am not giving a personal opinion, but the impartial evidence of an unprejudiced watch. The curves given in Figure 9.6 are for cats learning to escape from the box already described, whose door was held by a wooden button on the inside.

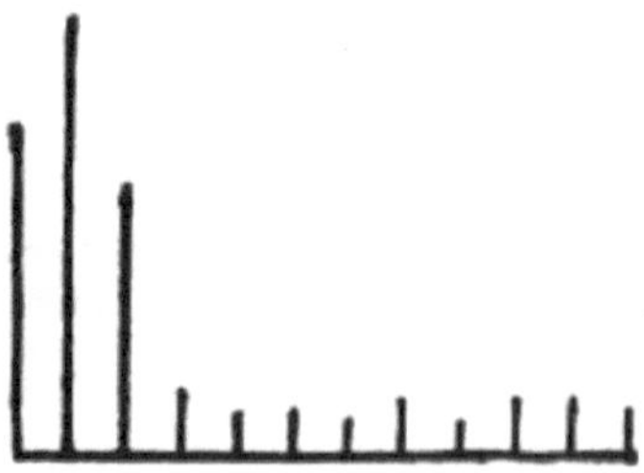

Figure 9.3

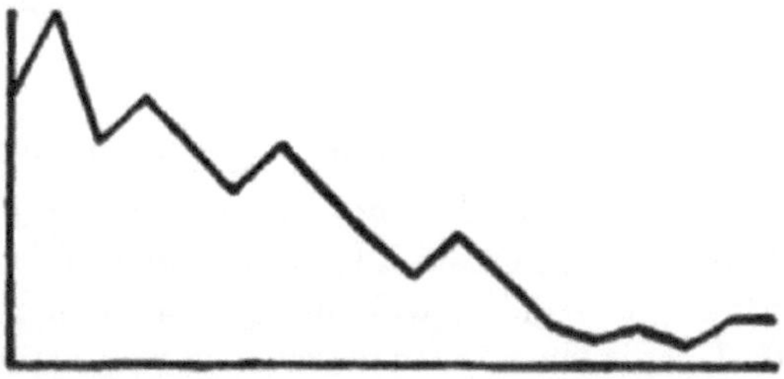

Figure 9.4

Some one may object that, true as all this may be, the intelligent acts reported of animals are in many cases such as could not have happened in this way by accident. These anecdotes of apparent comprehension and inference are really the only argument which the believers in reason have presented. Its whole substance vanishes if, as a matter of fact, animals can do these supposed intelligent acts in the course of instinctive struggling. They certainly can and do. I purposely chose, for experiments, two of the most intelligent performances described by Romanes in his Animal Intelligence – namely, the act of opening a door by depressing the thumbpiece of an ordinary thumb-latch and the opening of a window by turning a swivel (see pp. 420–422 and p. 425 of Animal Intelligence, by G. J. Romanes). Here I may quote from the detailed report of my experiments (Monograph Supplement to the Psychological Review, No. 8):

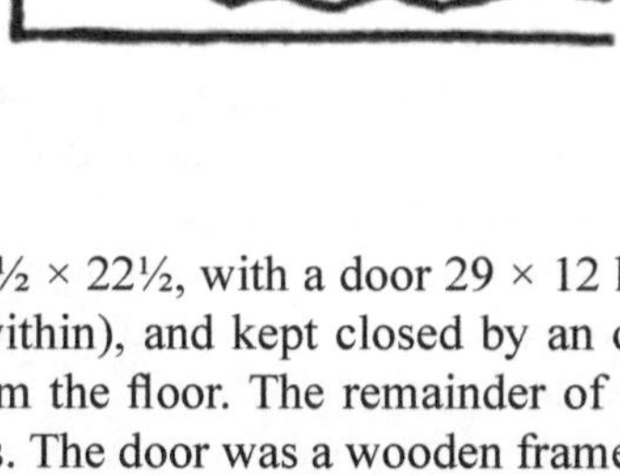

Figure 9.5

"G was a box 29 × 20½ × 22½, with a door 29 × 12 hinged on the left side of the box (looking from within), and kept closed by an ordiary [sic] thumb-latch placed fifteen inches from the floor. The remainder of the front of the box was closed in by wooden bars. The door was a wooden frame covered with screening. It was *not* arranged so as to open as soon as the latch was lifted, but required a force of four hundred grammes, even when applied to the best advantage. The bar of the thumb-latch, moreover, would fall back into place again unless the door were pushed out at least a little. Eight cats (Nos. 1, 2, 3, 4, 5, 6, 7, and 13) were, one at a time, left in this thumb-latch box. All exhibited the customary instinctive clawings and squeezings and bitings. Out of the eight, *all succeeded, in the course of their vigorous struggles, in pressing down the thumb-piece,* so that if the door had been free to swing open they could have escaped. Six succeeded in pushing both thumb-piece down and door out, so that the bar did not fall back into its place. Of these, *five succeeded in also later pushing the door open, so that they escaped and got the fish outside.* Of these, three, after about fifty trials, associated the complicated movements required with the sight of the interior of the box so firmly that they attacked the thumb-latch the moment they were put in."

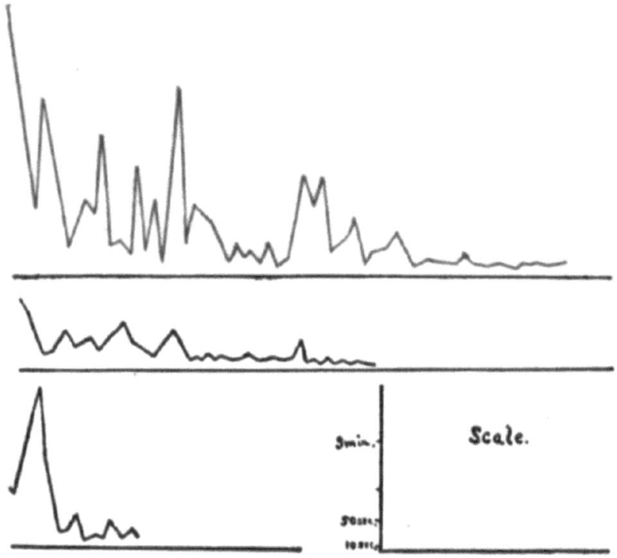

Figure 9.6

In the cases of No. 1 and No. 6 the combination of accidents required was enough to make their successes somewhat rare. Consequently weariness and failure offset the occasional pleasure of getting food, and after succeeding four and ten times respectively they never again succeeded, though given numerous opportunities. Their cases are almost a perfect proof of the claim that accident, not inference, makes animals open doors. For they hit upon the thing several times, but did not know enough to profit even by these experiences, and so failed to open the door the fifth and eleventh times.

Accident is equally capable of helping a cat escape from an inclosure whose door is held by a swivel.

"Out of six cats who were put in the box whose door opened by a button, *not one failed,* in the course of its impulsive activity, to push the button around. Sometimes it was clawed one side from below; sometimes vigorous pressure on the top turned it around; sometimes it was pushed up by the nose. No cat who was given repeated trials failed to form a perfect association between the sight of the interior of that box and the proper movements."

If, then, three cats out of eight can escape from a small box by accidentally operating a thumb-latch, one cat in a hundred may easily escape from a room by accident. If one hundred per cent of all cats are sure to sooner or later turn a button around when in a small box, one cat in a thousand may well escape from a room by accidentally turning a swivel around.

So far we have seen that when put in situations calculated to call forth any thinking powers which they possess, the animal's conduct still shows no signs of anything beyond the accidental formation of an association between the sight of the interior of the box and the impulse to a certain act, and the subsequent complete establishment of this association because of the power of pleasure to stamp in any process which leads to it. We have also seen that samples of the acts which have been supposed by advocates of the reason theory to require reasoning for their accomplishment turn out to be readily accomplished by the accidental success of instinctive impulses. The decision that animals do not possess the higher mental processes is re-enforced by several other lines of experiment – for example, by some experiments on imitation.

The details of these experiments I will not take the time to describe. Suffice it to say that cats and dogs were given a chance to see one of their fellows free himself from confinement and gain food by performing some simple act. In each case they were where they could see him do this from fifty to one hundred and fifty times, and did actually watch his actions closely from ten to forty times. After every ten chances to learn from seeing him, they were put into the same inclosure and observed carefully, in order to see whether they would, from having so often seen the act done, know enough to do it themselves, or at least to try to do it. In this they signally failed. Those who had failed previously to hit upon the thing accidentally never learned it later from seeing it done. Those who were given a chance to imitate acts which accident would sooner or later have taught them learned the acts no more quickly than if they had never seen the other animal do it the score or more of times. The animals, that is, could not master the simple inference that if, in a certain situation, that fellow-cat of mine performs a certain act and gets fish, I, in the same situation, may get fish by performing that act. They did not think enough to profit by the observation of their fellows, no matter how many chances for such observation were given them.

Equally corroborative of our first position are the results of still another set of experiments. Here the dogs and cats were put through the proper movement from twenty-five to one hundred times, being left in the box after every five or ten trials and watched to see if they would not be able at least to realize that the act which they had just been made to do and which had resulted in liberation and food was the proper act to be done. For instance, a dog would be put in a box the door of which would fall open when a loop of string hanging outside the box was clawed down an inch or so. Animals were taken who had, when left to themselves, failed to be led to this particular act by their general instinctive activities. After two minutes I would put in my arm, take the dog's paw, hold it out between the bars, and, inserting it in the loop, pull the loop down. The dog would of course then go out and eat the bit of meat. After repeating this ten times (in some cases five) I would put the dog in and leave him to his own devices. If, as was always the case, he failed in ten or twenty minutes to profit by my teaching I would take him out, but would not feed him. After a half hour or so I would recommence my attempts to show the dog what needed to be done. This would be kept up for two or three days, until he had shown his utter inability to get the notion of doing for himself what he had been made to do a hundred or more times. The mental process required here need not be so high a one as inference or reasoning, but surely any animal possessing those would, after seeing and feeling his paw pull a loop down a hundred times with such good results, have known enough to do it himself. None of my animals did know enough. Those who did not in ten or twelve trials hit upon an act by accident could never be taught that act by being put through it. And, as in the case of imitation, acts of such a sort as would be surely learned by virtue of accidental success were not learned a whit sooner or more easily when I thus showed them to the animal.

An interesting supplement to these facts is found in the following answers to some questions which I sent to the trainer of one of the most remarkable trick-performing horses now exhibited on the stage. The counting tricks done by this horse had been quoted to me by a friend as impossible of explanation unless the horse could be educated by being put through the right number of movements in connection with the different signals.

Question 1. – If you wished to teach a horse to tap seven times with his hoof when you asked him "How many days are there in a week?" would you teach him by taking his leg and making him go through the motions?

Answer. – "No!"

Question 2. – Do you think you *could* teach him that way, even if naturally you would take some other way?

Answer. – "I do not think I could."

Question 3. – How would you teach him?

Answer. – "You put Figure 9.2 on the blackboard and *touch him on the leg* twice with a cane, and so on."

The counting tricks of trained horses seem to us marvelous because we are not acquainted with the simple but important fact that a horse instinctively raises his hoof when one pricks or taps his leg in a certain place. Just as once given, the cat's instinct to claw, squeeze, etc., you can readily get a cat to open doors by working latches or turning buttons, so, once given this simple reflex of raising the hoof, you can, by ingenuity and patience, get a horse to do almost any number of counting tricks.

Probably any one who still feels confident that animals reason will not be shaken by any further evidence. Still, it will pay any one who cares to make scientific his notions about animal consciousness to notice the results of two sets of experiments not yet mentioned. The first set was concerned with the way animals learn to perform a compound act. Boxes were arranged so that two or three different things had to be done before the door would fall open. For instance, in one case the cat or dog had to step on a platform, reach up between the bars over the top of the box and claw down a string running across them, and finally push its paw out beside the door to claw down a bar which held it.

The animal's instinctive impulses do often lead it to accidentally perform these several acts one after another, and repeated accidental successes do in some of these eases cause the acts to be done at last in fairly quick succession. But we see clearly that the acts are not thought about or done with anything like a rational comprehension of the situation, for the time taken to learn the thing is much longer than all three elements would take if tackled separately; and even after the animal has reached a minimum time in doing the acts, he does not do the things in the same order, and often repeats one of the acts over and over again, though it has already attained its end.

The second set comprised experiments on the so-called "memory" of animals. I will describe only one out of many which agree with it. A kitten had been trained to the habit of climbing the wire-netting front of its cage whenever I approached. I then trained her to climb up at the words "I must feed those cats." This was done by uttering them and then in ten seconds going up to the cage and holding a bit of fish to her at its top. After this had been done about forty times she reached a point where she would climb up at the signal about fifty per cent of the times. I then introduced a new element by sometimes saying, "I must feed those cats," as before, and feeding her, and at other times saying, "I will not feed them," and remaining still in my chair. At first the kitten felt no difference, and would climb up just as often at the wrong signal as at the right. But gradually (it took about four hundred and fifty trials) the failure to get any pleasure from the act of climbing up at the wrong signal stamped out the impulse to do so, while the pleasure sequent upon the act of climbing up at the other signal made that her invariable response to it. Here, as elsewhere, the absence of reason was shown by the cat's failure at any point in these hundreds of trials to think about the matter, and make the easy inference that one set of sounds meant food, while the other did not. But still better proof appears in what is to follow. After an interval of eighty days I tried her again to see how permanent the association between the signal and act was. It was permanent to the extent that what took three hundred and eighty trials before took only fifty this time, for after fifty trials with the "I will not feed them" signal, mixed up with a lot of the other, the cat once more attained perfect discrimination. But it was not permanent in the sense that the cat at the first or tenth or twentieth trial felt, as a remembering, reasoning consciousness surely ought to feel, "Why, that lot of sounds means that he won't come up with fish." For instead of at first forgetting and for a while climbing up at the *I will not feed them,* and then remembering its previous experience and at once stopping the performance it had before learned was useless, the cat simply went through the same gradual decreasing of the percentage of wrong responses until finally it always responded rightly.

What has so far been said is true regardless of any prejudice or incompetence on my part, for the proof in all cases rests not on my observation, but on impartial time records or such matters of fact as the escape or nonescape, the climbing or not-climbing of the animals. I may add that in a life among these animals of six months for from four to eight hours a day I never saw any acts which even *seemed* to show reasoning powers, and did see numerous acts unmentioned here which pointed clearly to their absence.

All that is left for the fond owner of a supposedly rational animal to say is that though the average animal, the typical dog or cat, is by these experiments shown to be devoid of reasoning power, yet *his* dog or *her* cat is far above the average level, and is therefore to be judged by itself. He may claim that just because my average animals failed to infer, we have no right to deny inference to all, particularly to his. Is it not fair to ask such a one to repeat my experiments with his supposedly superior animal? Until he does and systematically tries to find out how its mind works and what it is capable of, has he any right to bear witness? It may also be said that of the number of people who witnessed the performances of my animals after they had fully learned a lot of these acts, but had not seen the method of acquisition, all unanimously wondered at their wonderful intellectual powers. "How *do* you teach them?" "Where did you get such bright animals?" "I always thought animals could think," and such like were common expressions of my visitors. The fact was that the dogs and cats were picked up in the street at random, and that no one of them had thought out one jot or tittle of the things he had learned to do. The specious appearance of reasoning in a completely formed habit does not involve the presence or assistance of reasoning in the formation of the habit.

Here, at the close of this account, I may signify my willingness to reply, so far as is possible, to any letters from readers of the Popular Science Monthly who may care to ask questions about any feature of animal intelligence.

9.2 CORRESPONDENCE.

"Do Animals Reason?"

Dr. Edward Thorndike's interesting account, in our August number, of his investigations touching the reasoning power of animals has brought us a large number of letters questioning some of the main conclusions set forth in the article, and criticising the method of the inquiry. Not having room for all these communications, we print one of them, and add extracts from two others. These represent the principal objections urged by the various writers against the conclusions drawn by the author of the article from his experiments.

Editor Popular Science Monthly:

SIR: The first reading of Dr. Thorndike's article Do Animals Reason? in the August Popular Science Monthly, gave the impression, which has been deepened by subsequent perusal, that his experiments were not only inadequate to solve the question, but unfairly chosen.

A dog or a cat, utterly hungry, is placed in a box, from which it can escape "by performing some simple (?) action, such as pulling a wire loop, stepping on a platform or lever, clawing down a string, or turning a wooden button."

In the first place, what tends to destroy the reasoning power more than litter hunger? This intense physical craving begets frenzy rather than reason. The more intense this primeval desire, the greater the demand upon primitive instinct for its satisfaction. In the open the cat will jump at a bird, the dog at a bone. If the bird be up a tree, the cat will climb; if the bone be buried, the dog will burrow. Climbing and burrowing are deep-rooted developments of the feline and the canine nature.

Put a dog or a cat, utterly hungry, in a box and hang a piece of meat outside. Instinct prompts a jump through the bars of the box at the meat, and the greater the number of unsuccessful attempts the less the likelihood of the animal with a gnawing stomach sitting down to scrutinize the mechanical construction of the box to the point of perceiving that by stepping on a lever it will open a door. How many millions of years did it take two-legged man to arrive at the perception of the use of the lever? Did the shaggy biped arrive at that perception by sitting down when utterly hungry and looking at a lever; or did he, through countless generations, by some such chance as lifting a stone with a stick, come to the knowledge of weight and fulcrum?

Put an anthropoid ape, some several degrees nearer man in intelligence than a cat, in a modern office elevator that moves by the push of an electric button, suspend the elevator between two stories, and what do you suppose that anthropoid ape will do?

Put a schoolgirl fresh from belles-lettres and matinées in the cab of a locomotive and tell her to run it to the next station. She can not but know that steam will make the wheels go round, but what will she do in the maze of throttles, handles, disks, and rods that confronts her? What will she do if utterly hungry?

Take a laborer from his pick and shovel on the railway embankment and put him at the desk of the general manager. He can read and write. Let the messenger boys and clerks shower him with the letters and telegrams that bombard that desk every day, and let him try to settle the questions to which they give rise.

Now, why can not the schoolgirl run the locomotive, the laborer the railroad? Because the relations of things necessary to the tasks have never been imprinted upon their registering cells; because, in the latter case at least, of the lack of power of co-ordination – that is, the lack of the power of abstract reasoning that the task involves.

Why can not anybody do anything as well as anybody else? Because certain relations have been more deeply impressed upon certain brains than upon others;

because of the greater power of certain brains to co-ordinate certain relations, their greater ability to give concrete manifestation of the result of such co-ordination through the efferent nerves. Otherwise any one of us could design a bridge, compose a symphony, or organize a trust.

The oftener relations are impressed upon the registering cells, the more readily are those relations co-ordinated, provided the brain structure be of the requisite caliber. Reiterated impression through the ages of the relations between their needs and surrounding things, together with the development of structural capacity, has led the beaver to build his dam, the bee the honeycomb, the ant its village, the bird its nest. In each case the registered impressions have led to notion made possible by long-continued contact between structure and environment; the actions are the result of development that has proceeded mite by mite through unknown time. The brain of neither bird nor beast nor man will immediately co-ordinate radically new impressions received in a radically new environment into coherent action that leads to definite result.

Here is an example within the writer's immediate knowledge: At the age of seventeen a boy entered the service of one of the large railway systems as a clerk in the passenger department. Through eleven years of enthusiastic and concentrated endeavor to master the details of the service he rose to the head of the clerical force – that is, the reiterated impression upon his brain cells of the functions of the passenger service led to that co-ordination which resulted in efficient action. Then he became employed in the office of a large coal-mining company. For several days it was with the utmost difficulty that he could bring his attention to bear upon the new tasks. While seated at the desk in the coal office the old railway problems would chase through his mind; when he began to write the initials of the Pittsburgh Consolidated Coal Company, he would find that he had written the initials of the Pittsburg, Cincinnati, Chicago and St. Louis Railway Company; instead of the initials of the Pittsburgh, Fairport and Northwestern Dock Company, the initials of the Pittsburgh, Fort Wayne and Chicago Railway Company. The latter initials in each case would appear upon the paper before he knew it, actually without his knowing that he had written them. The entirely unfamiliar routine entailed by the custody of bank accounts, coal leases, deeds and contracts, reports of coal shipments, and the handling of vouchers, became adjusted in his brain bit by bit through many weeks, and it was months before he could co-ordinate the new impressions into broad and well-defined reasoning. If he had been utterly hungry through all the period of the new service, it might have taken years.

Now, what can be expected of a dog or a cat, whose mental processes have been adjusted by inheritance and experience to life in the fields and jungles, when placed in a box, utterly hungry, to study mechanical contrivances? It is manifest that if the brain of a dog or a cat would become adjusted to the radically unfamiliar steps necessary to release it from such a radically unfamiliar environment, that adjustment could only come by extremely slow degrees. Voluntary perception is almost beyond the limits of expectation, and the leading of the animal through the necessary steps would have to be repeated time after time before the impressions upon its brain would reach any degree of permanence, especially as its brain would be lacking in attention, and the repeated handling be an annoyance to it. But that by such tutelage the animals, or a proportion of them, arrived at a knowledge of the means necessary to escape from the box is shown by Dr. Thorndike himself. "If one repeats the process, keeps putting the cat back into the box after each success, the amount of useless action gradually decreases, the right movement is made sooner and sooner, until finally it is done as soon as the cat is put in." But he says: "This sort of a history is not the history of a reasoning animal. It is the history of an animal who meets a certain situation with a lot of instinctive acts. . . . Little by little the one act becomes more and more likely to be done in that situation, while the others slowly vanish. This history represents the wearing smooth of a path in the brain, not the decisions of a rational consciousness."

Wherein, however, does this differ from the manner in which hundreds of clerks in offices finally learn routine work and mechnically go through the motions necessary to its performance? Do not the actions of thousands of laborers in field and factory seem to proceed from a wearing smooth of a path in the brain, rather than from rational consciousness? Yet they can not be said to be devoid of reason. Is not a great proportion of the daily actions of any one of us gone through from force of habit, almost by instinct?

The word reason does not apply alone to the mental processes of a Helmholtz,[2] but to the co-ordination, however slight, of relations that result in definite action even of a humble organism. Herbert Spencer has clearly shown that instinct and reason differ in degree and not in kind.

Dr. Thorndike lays stress upon the fact that a "cat which, when first put in, took sixty seconds to get out, in the second trial eighty, in the third fifty, in the fourth sixty, in the fifth fifty, in the sixth forty," etc., and remarks: "Suppose the cat had, after the third accidental success, been able to reason? She would then have, the next time and all succeeding times, performed the act as soon as put in." Not long ago the writer and a man whose high intelligence can not be questioned, in moments of relaxation were trying to do one of the familiar ring puzzles – endeavoring to separate a ring from two others of peculiar shape and then to join the three. After repeated trials, one would loosen it, but could not replace it; the other finally succeeded in replacing it, but could not loosen it. Then the one could replace it, but not loosen it; the other loosen, but not replace it, and each was closely watching the other all the time. It was half an hour or more before either could both loosen and replace the ring, occasional successful attempts not being repeated until after several succeeding failures. Contrast the relation of the brain of the dazed and indifferent and peculiarly bedeviled cat to the puzzle presented to it by the inside of the box with the earnest effort of the two men to solve the ring puzzle. Who has not found a task more difficult the fifth or sixth time than the second or third, and has only performed it with ease after repeated attempts of varying degrees of success and failure?

In conclusion, the writer begs leave to relate an incident, which has not before appeared in print, that profoundly impressed him with the belief that at least in one instance one particular animal displayed reason. One Sunday morning, a dozen years or more ago, he was standing on the bank of the Ohio River at the Sewickley Ferry. A family group, accompanied by a large Newfoundland dog, hailed the ferryman and got in his boat, leaving the dog, which persuasively barked and wagged his tail, on the bank. As the boat pulled out into the stream the dog whined, and then made ready to leap in after it. Then he stopped at the water's edge, and, with head down, gazed intently at the river for several seconds – it seemed a minute or more. Then he ran up the bank more than a hundred feet, stopped, looked at the receding boat, plunged into the stream, and swam vigorously. The current, bearing him down, made his course diagonal to the bank. A boy standing by my side said: "Isn't that a smart dog? If he'd been a crazy dog he'd have jumped in where he was, but he ran up the bank so the current wouldn't wash him down away from the boat."

But the dog, swimming with all his vigor, was borne past the boat when within twenty feet or so of it; he endeavored to straighten his course without success, and then, in a long semicircle, swam around to the near bank, landing two or three hundred feet below the place whence the ferryboat had started.

What this dog would have done if placed, utterly hungry, in a box from which he could only liberate Himself by stepping on a platform or tinning a wooden button, I do not know.

LOGAN G. MCPHERSON.[3]
PITTSBURGH, *August 3, 1899.*

MR. FREDERIC D. BOND, of 413 South Forty-fourth Street, Philadelphia, writes: Of the accuracy of Dr. Thorndike's experiments I have no doubt, but certain facts connected with them seem to deprive the observations of much of their relevance.

Dr. Thorndike states that he arranged his experiments to give reasoning every chance to display itself, if it existed, and to observe those in which the acts required and the thinking involved were not far removed from the acts and feelings of ordinary animal life. Of these experiments one of the chief was to determine whether and in what way a cat would escape from a box opening by turning a button. Now, I submit that in this and the succeeding experiments the conditions Dr. Thorndike fancied to exist by no means did so. Simple as the release of a door by a button seems to us, the apparent simplicity arises merely from our empirical knowledge of what does happen in such a situation. Actually to think out the rationale of the matter, as an animal having no experience either personally or from heredity would have to do, involves very complex mental processes. The environment of a human being is vastly different from an animal's, though of this fact we constantly lose sight in reasoning; of mechanical appliances and principles, for example, an animal knows nothing, and yet we are too apt to suppose it regarding the world with a store of ancestral and individual experiences utterly foreign to it; and then, on its failing to do what, in the light of such experience, seems to us easy, we proceed to call into question its possession of reason. . . .

That the cats did finally learn to escape shows, according to Dr. Thorndike, "the wearing smooth of a path in the brain, not the decisions of a rational consciousness." May I ask Dr. Thorndike what possible reason could a cat have to suppose that what happened once must needs happen again? Does Dr. Thorndike fancy his own knowledge of a million like matters was acquired by reason, and not empirically elaborated by processes of exactly the same sort as the cats went through? Let this experiment be tried on a healthy infant of two years, and I am of the opinion that the results would be the same as with the cat; yet the infant undoubtedly carries on "thinking processes similar, at least in kind, to our own," which Dr. Thorndike implicitly denies to his cats.

The chief cause of the inability of students to reach concordant results in this matter of animal intelligence appears to lie in a certain uncritical assumption often made. That all consciousnesses have a certain field of presentations, that to this field they attain, that because of it they feel and will, are fundamental facts; but the belief that attention or feeling or will differs *per se* in different consciousnesses, other than as the field to which they are at the moment related, differs – this is an utterly unwarranted assumption. According to the action of its environment, each conscious being must know the world just so far as is needed to conform its existence thereto, or else it must perish; but whether such knowledge, which is acquired by experience only, be quite small, as with animals, or somewhat larger, as with man, there is no reason to suppose that the attention, feeling, or will of the animal differs in itself from the same psychological state in man.

MR. ANDREW VAN BIBBER, of Cincinnati, Ohio, says: Animals, and especially wild ones, have no bank account or reserve, and have to face new conditions daily, and yet they make a living where man would starve.

When I was out in Colorado and Utah, years ago, I used to know of animals removing the bait nicely from dangerous traps without springing the trap. I knew of a dog who went over a mile to call his owner to the aid of a boy who had broken his leg, and who would not be refused till understood. This is brutish "instinct," is it? something that Dr. Thorndike can't define. Will "instinct" teach a tired, half-starved horse to eat oats if you set them before him? Dr. Thorndike would say "Yes," but Dr. Thorndike would be wrong unless that horse knew from personal past experience what oats were. What animals learn (like the human animal) they learn chiefly by experience. They accumulate facts in their minds and use them.

I served in the cavalry of the Armies of the Tennessee and the Cumberland,[4] and I know that instinct will *not* cause a hungry horse to touch oats unless he knows from his own experience what oats are. We used to capture horses in Mississippi which had never seen oats. It is all corn down there. We would bring them into camp tired out and hungry, and would pour out our oats for them. Not one of them would touch the oats. You could leave the hungry horses hitched for twenty-four hours before oats, and not one grain would they touch. They would stand there and starve. We had to throw up their heads and fill their mouths full of oats. If we stopped there, they would spit them out. We had to grab their jaws and work them sideways until they had a good taste. Then they understood, and ate oats right along. Plenty of such horses in Mississippi to-day. . . .

If Dr. Thorndike tried his intelligent "Experiment No. 11" with a two-year-old cat, why didn't he try it with a two-year-old human? I guess he would have found an equal amount of ignorance of the mechanism of door fastenings, which comes only with teaching, and would have produced only struggles and screaming.

Editorial Endnotes

1 *Lloyd Morgan*: Conwy Lloyd Morgan (1852–1936) was a British psychologist and ethologist known for "Morgan's Canon," a psychological axiom which holds that when interpretating animal behavior scientists should aim for the simplest possible explanation of said behavior (also see M. F. Washburn). Although often interpreted this way, Morgan did not introduce this axiom in opposition to anthropomorphism, nor did he insist that simplest explanation was always the best one.

2 *Helmholtz*: Hermann von Helmholtz (1821–1894) was a German physicist whose contributions span several areas, including the physiology and physics of human vision, acoustics, electromagnetism, and thermodynamics.

3 Logan G. McPherson: Logan Grant McPherson (1863–1925) was an American economist who worked in various positions in the coal and railroad industries and lectured on transportation at Johns Hopkins University.

4 *Armies of the Tennessee and the Cumberland*: two Union armies operating in the western theater of the American Civil War. The Army of the Tennessee – not to be confused with the Confederate Army of Tennessee – was initially commanded by military officer and later U.S. president Ulysses S. Grant, who was promoted to commander of all Union armies in the final years of the war.

Editorial Headnote

10. Ernest Ingersoll, "Do Animals Commit Suicide? A Study of Brute Limitations," from *The Wit of the Wild.* New York: Dodd, Mead and Company, 1906.

Ernest Ingersoll (1852–1946) was a naturalist, journalist, and author of numerous nature and animal books. Beginning with his participation as a zoologist in an 1874 survey expedition led by geologist Ferdinand Vandeveer Hayden and throughout the 1870s and 1880s, Ingersoll spent a significant amount of time traveling the American West, the subject of various articles published in magazines like *Harper's*, *Scribner's*, and the hunting and outdoor magazine *Forest and Stream*, for which he worked as natural history editor. From the late 1880s into the 1920s, Ingersoll published several books on wild animal life – including *The Wit of the Wild*, from which the following chapter is taken – that combined reprinted material from his magazine contributions with original essays.[1]

In the following chapter, Ingersoll discusses depictions of animal suicide in a number of contemporary animal stories and anecdotes, taking as a point of departure the story of "The Springfield Fox" from Ernest Thompson Seton's *Wild Animals I Have Known* (1898), a book which by then had already become a prominent target of criticism in the context of the "nature fakers" controversy discussed in the introduction to the present volume. Ingersoll's books were among those praised as examples of the "right" kind of nature writing by naturalist John Burroughs in his March 1903 *Atlantic Monthly* essay on "Real and Sham Natural History" – infused with a love for nature yet unswerving in their commitment to scientific accuracy and immune to the distortions of "sentiment." While Ingersoll was less outspoken in his nature fakery accusations than Burroughs, he, too, criticised the work of writers like Seton and Connecticut minister William Joseph Long, whose animal stories were used in contemporary schoolbooks.[2]

The topic of animal suicide ties in with, and sits at the extreme end of, the debates around the awareness of death in animals, with the oft-asserted absence of the latter figuring as yet another entry in the arsenal of human exceptionalist claims.[3] Ingersoll would likely have agreed with Nathaniel Southgate Shaler's (see N. S. Shaler, Vol. 2, 3) claim that what separated humans from animals was the human self's conscious confrontation with the problem of finitude rather than the instinctual animal fear of death, which, although crucial to self-preservation, ultimately remained "blind and unconscious of its object."[4]

Notes

1 Clyde L. MacKenzie, "Biographic Memoir of Ernest Ingersoll: Naturalist, Shellfish Scientist, and Author," *Marine Fisheries Review* 53, no. 3 (1991): 23–29.

2 Ralph H. Lutts, *The Nature Fakers: Wildlife, Science and Sentiment* (Charlottesville: University of Virginia Press, 2001), 38.

3 For discussions of this question, see John M. Marzluff and Tony Angell, *In the Company of Crows and Ravens* (New Haven: Yale University Press, 2005); Barbara J. King, *How Animals Grieve* (Chicago: University of Chicago Press, 2014); André Gonçalves and Susana Carvalho, "Death Among Primates: A Critical Review of Non-Human Primate Interactions Towards Their Dead and Dying," *Biological Reviews* 94, no. 4 (2019): 1502–29.

4 Nathaniel S. Shaler, *The Individual: A Study of Life and Death* (New York: D. Appleton and Company, 1900), 200.

10

DO ANIMALS COMMIT SUICIDE? A STUDY OF BRUTE LIMITATIONS (1906)

Ernest Ingersoll

Source: Ernest Ingersoll, *The Wit of the Wild* (New York: Dodd, Mead and Company, 1906), 196–210.

IN that fascinating book, "Wild Animals I Have Known," no chapter is more alluring in its mixture of plain fact and fanciful interpretation than the one which chronicles the doings of the Springfield fox and its dramatic end. This story, which purports to be that of the real field-life of an actual American red fox, is made vivid by the human perceptions and sentiments attributed to the animal as its own, among which is a perfect comprehension of death. This fox, it appears, knew thoroughly the use and efficacy of poison. That such an animal may learn and teach its young to detect by smell the taint of poison in a piece of meat, and so refuse to eat the morsel on the broad ground that it is unsafe, or on the narrower ground that it has something to do with the enemy, man, all of whose works are to be suspected, is easy of belief. "Vix knew right well," says Mr. Seton, "what poisoned bait was; she passed them by or treated them with contempt, but one she dropped down the hole of an old enemy, a skunk, who was never afterward seen."

One must wish this naturalist would substantiate by more particulars this revengeful murder by indirect means, which seems to me to imply more, in several directions, than we have been able hitherto to grant to the intelligence of even a fox; yet this is easy of acceptance beside the further deeds of this prodigy, after its mate had been shot and its home ravaged by men and dogs. Three of its cubs are killed and a fourth is made captive and chained in the farmer's yard. Night after night, urged by mother-love, old Vixen comes to feed her imprisoned bairn, and to try by every strength and device of tooth and nail and mind to get it free. At last she becomes convinced that no effort of hers can loosen the chain, and the next night she turns to her last resort:

> "Like a shadow she came, and in a moment was gone, and Tip seized on something dropped, and crunched and chewed with relish what she brought. But even as he ate, a knife-like pang shot through, and a scream

DOI: 10.4324/9781003032717-11

> of pain escaped him. Then there was a momentary struggle and the little fox was dead.
>
> "The mother's love was strong in Vix, but a higher thought was stronger. She knew right well the poison's power, . . . now at last when she must choose for him a wretched prisoner's life or sudden death, she quenched the mother in her breast and freed him by the one remaining door."
>
> After that Vixen is no longer seen in the neighborhood, – "gone, maybe, deliberately," says the author, "from the scene of a sorrowful life, as many a wildwood mother has gone, by the means that she herself had used to free her young one – the last of all her brood."

Now, here is an assertion, as of observed facts; first, of two premeditated murders by a brute animal, one from motives of prudence and revenge, and another from motives of the highest moral import; and, second, an implication, amounting almost to an assertion, of conscious suicide. If it were no more than a bit of pathetic fancy it would be worth notice only as evidence of a popular notion, but it is woven into what is declared to be a true story of the conduct of life by an actual fox, and thus takes its place as an assertion of history.

But do animals ever kill one another, except by accident, in anger or as prey? And, do animals ever commit suicide?

What is suicide? Literally, and taking the word merely as it stands, nothing more than self-killing; but the customary and now only proper definition involves the idea of intentional, voluntary self-destruction, and this implies on the part of the being so acting an understanding of the circumstances of life and death, – at least of the difference between them.

The literature of dog-stories – if one may use the expression – abounds in anecdotes of pets, or of canine servants of the hunter and shepherd, which have grieved themselves into a speedy decline and death by exhaustion due to refusal of food, on the graves of their dead masters; which have refused to leave them in moments of great peril, as in a shipwreck, and so have perished; or which in some other way have sacrificed their lives for their friends or at least with them. All these incidents, however, only illustrate that remarkable dependence and sense of duty which distinguish the dog – an inheritance from innumerable generations of trained ancestors taught to be "faithful unto death."[1]

Something of the same disposition is occasionally shown in the horse, which stands next to the dog in long association with man. The sense of dependence upon and affection for a master is so exceedingly strong in most dogs, that when one of them is suddenly deprived of companionship it is no wonder it places itself as near as possible to where its benefactor has been laid, and waits for his reappearance, fearing to leave the place even for a moment, lest it shall miss his return. This very attitude proves that it does not apprehend the finality of its master's condition. At last exhaustion, and the nervous depression resulting from grief, mysterious dread and disappointment, causes mortal illness, – but there is no evidence the dog foresees or intends a fatal result. In the other class of cases, the idea of remaining with a master, danger or no danger, or of defending him, as if it were itself attacked, is so strong in a good dog as to overcome timidity and prudence.

Something different from this is needed as evidence of intentional suicide; and when the incidents alleged to furnish such further evidence are examined, they are always, so far as has come to my knowledge, found wanting. Here, for an instance, is one published in *Notes and Queries*[2] during 1898. An English gentleman, who owned a small terrier, was obliged to go to the Continent for his health, and after a few weeks died. When the news of his death came the dog seemed to understand what had happened, "and shared the grief of the family to such an overwhelming extent that one day it went to an upper window and jumped out, killing itself in a very distressing way." Any one considering this narrative a moment may see how many assumptions it contains, and especially how suppostitious is the conclusion. There is no evidence that it was not an accident. Terriers often fall from windows and sometimes jump from them. Only a short time ago a dog sprang from a lofty window-sill in New York in an attempt to catch a bird; it either did not know, or more probably forgot, the peril of the leap. My own dog, the first time it was taken in a boat upon a lake, stood on the prow of the boat a while and then deliberately sprang overboard, where it was immensely surprised and alarmed to find itself struggling in deep water: all the water it knew about previously was very shallow.

Had not the circumstances of both these last-mentioned cases been known, and especially had they been associated with deaths in the families, or something else remarkable, they might well have been adduced as examples of conscious self-destruction.

A correspondent of *The Field*[3] some time ago gave a long account of how a terrier between repeated attacks of "fits," first dashed himself off a high wall and then ran into the sea, where, after a few half-hearted strokes, it turned on its back and drowned. The observer declares his belief that the animal deliberately put itself out of its miseries, but most of us cannot but attribute the acts to temporary insanity. It is difficult to separate sympathy, romance and tradition from facts and cool judgment, in such cases. It has been an ancient belief, for example, coming down from the middle ages, that a scorpion put within a circle of fire will sting itself and die as soon as it perceives that there is no escape; but not only do not modern scorpions turn to the *felo de se*[4] as a release from expected pain, but no scorpion could sting itself to death if it tried.

Perhaps the influence of these old fancies lingered in the mind of Mr. Seton when in another chapter of the same book he suggests suicide as the final proud and praiseworthy act of a wild stallion, which for years had baffled all pursuers. At last, however, an organized effort for the capture of the "pacing mustang" was at the point of success:

> "The man rejoiced, but the mustang gathered his remaining strength for one more desperate dash. Up, up the grassy slope from the trail he went, defied the swinging, slashing rope, and the gunshot fired in air, in vain attempt to

> turn his frenzied course. Up, up and on, above the sheerest cliff he dashed, then sprang away into the vacant air, down – down – two hundred feet to fall, and landed upon the rocks below, a lifeless wreck – but free."

All equine animals are subject to insane panic, when they lose all self-control and may rush blindly to accidental destruction; but this author, who claims to be relating actual facts, doesn't mean us to understand this incident in that way. I should like to hear his explanation of how this black mustang, or how Vixen, the fox-mother, had arrived at the knowledge that death meant freedom, and was preferable to the halter? What experience had either of captivity? No such condition exists in nature: what reason had they to foresee, much less dread it? Upon what data could they create a mental picture (granting imaginative ability) calculated to alarm them so profoundly?

If horses or foxes comprehend so much as that, why do not thousands of overworked, half-fed, cruelly abused domestic animals, and many a starving or tortured wild one, commit suicide every day? Why do not fur-bearing animals caught in traps kill themselves at once instead of dying by inches, or merely gnawing off the fixed foot, – an act, in my opinion, due to an effort to relieve the dreadful pain, and not to a deliberate method of release. Either they do not know enough, or else their sense of moral responsibility is superior to that of thousands of their masters.

So enamored is Mr. Seton of this conceit of suicide among wild creatures that he resorts to it again as the climax of his "Biography of a Grizzly," – a childish performance, at best, for a man who has shown such literary ability. He disposes of his hero, supposed to be a typical bear, living an average life, when old age and rheumatism make him feel ill and unfit, by making him go to a certain narrow valley which he (the bear) knows of, and which is strewn with the remains of animals that have perished in the fumes of a stream of noxious gas that flows out of a hole in the rocks. The disgusting odors of Death Gulch[5] "had a message for him," and Bruin meditates that it is far to Yellowstone Park, where he may end his days in peace under the protection of Uncle Sam's hospitality – though how could a grizzly have learned that? Moreover, as the sage animal reflects, "What's the use?"

"Here in this little garden was all he sought; here were peace and painless sleep: He knew it; for his nose, his never-erring nose, said 'Here! here! now!' He paused a moment at the gate, and as he stood the wind-borne fumes began their subtle work. . . . A rush of his ancient courage surged in the grizzly's rugged breast. He turned aside into the little gulch. The deadly vapors entered in, filled his huge chest, and tingled in his vast, heroic limbs, as he calmly lay down on the rocky, herbless floor, and gently went to sleep."

I can leave to others the literary question whether it is good to wind up a story, alleged to be of facts, with a purely imaginative dénouement. My contention now is that animals do not know anything whatever about life and death as contrasted or correlative conditions; and can have no idea that life may be ended, or that death is an alternative state which may be arrived at by fatal means.

Young beasts do not recognize death when they see it manifested in a lifeless body – though the highest apes seem to have some glimmering of the truth – but will linger about a mother that has been shot and try to awaken her attention. Older animals usually recognize a dead body as dead, but the state seems to mean to them only a mysterious disability, – incapacity for resistance and readiness to be eaten at leisure. In the case of carnivores, the last is probably the most vivid impression, and many of them will devour almost at once a partner, or even mate or offspring, killed by their side, when not themselves too much alarmed to take advantage of the lucky provision. Hunters constantly meet with instances of this "cannibalism."

I have watched with interest the behavior of my dogs toward dead animals. They would approach them cautiously, uncertain whether there were not some trick of play or hostile ruse in the immobility, until they could smell the cold form, then would turn away with an expression of wonder and disgust, but no further interest. Hunting animals have learned that in order to feed upon their prey they must reduce it to complete disability: it is the submission – not the *death* – of the creature which they seek when they strike. Such was the idea in the minds of the retrievers, made much of by Romanes, which kill one of two wounded ducks when they find it impracticable to retrieve both birds alive, as they are expected to do. The dog uses its natural dog sense in completing the disablement of an unmanageable thing, in order to accomplish that part of its mission which it feels of the highest importance, viz.: To get the game ashore, somehow. The retriever does not seek the *death* of the duck, *per se,* but merely its instantaneous acquiescence in his plan.

How could a dog or any other brute creature know of death apart from its outward aspects of disability and subsequent dissolution? What more, really, do *we* know about it, aside from our belief in Divine Revelation, or in the deductions of metaphysics? What data have the brutes for supposing that it gives "surcease of sorrow,"[6] or offers any refuge from distress, or even that such a change can be obtained by one's own act?

To comprehend the *fact,* not to say the nature, of death, one must comprehend the fact of self-life, and all that we know of the range of brute intelligence leads us to deny its ability to postulate self-existence. No experience can avail brutes in judging the effect of being lifeless, and every case of death seen must seem to the onlookers (if they "sense" it at all) utter ruin – something to be strenuously avoided. This is the natural physical view of death which must prevail throughout all nature, or life would come to an end. Everything in the natural world shapes itself and tends toward the *preservation,* in order to insure the *propagation,* of life. All feral instincts face that way, and to impartial laws and processes, with which individuals have nothing to do, is alone intrusted the duty of keeping within bounds the degree of multiplication.

Self-sacrifice is a supernatural, human idea, utterly at variance with all nature not human, and subversive of its prime reason and motive for existing. Nothing could be more illogical, or wicked, in the brute world, than the possibility of intentional self-destruction. Dogs and other animals may sometimes recklessly expose their lives to fatal peril, or, influenced by some nervous stress or extreme emotion, fall ill and die, or do blindly some fatal act; but we may be sure that no animal ever truly commits suicide, because no animal has any conception that it is possible to do so.

Editorial Endnotes

1 *"faithful unto death"*: from Revelation 2:10: "be thou faithful unto death, and I will give thee a crown of life" (KJV).
2 Notes and Queries: a British weekly founded in 1849 as a correspondence magazine that provided a forum of exchange between academic specialists and laypeople.
3 The Field: a British monthly founded in 1853 which featured articles on country life and outdoor activities like hunting or fishing.
4 felo de se: "felon of him-/herself," a legal term used in reference to a person who died by suicide in a context where suicide was criminalized, with implications for the treatment of the deceased's estate and form of burial.
5 *Death Gulch*: a geothermal site in what is now Yellowstone National Park, Wyoming. The site was "discovered" in 1888 by geologist Walter Harvey Weed. Its name is derived from the emission of hydrogen sulfide and other toxic gases responsible for the demise of various animals.
6 *"surcease of sorrow"*: from Edgar Allen Poe's poem "The Raven" (1845): "Eagerly I wished the morrow; – vainly I had sought to borrow/From my books surcease of sorrow – sorrow for the lost Lenore."

Editorial Headnote

11. William Temple Hornaday "The Psychology of Wild Animals," *McClure's Magazine* 30, no. 4 (February 1908): 469–79.

William Temple Hornaday (1854–1947) was a zoologist, taxidermist, and conservationist. His interest in taxidermy began during his time working for the museum of Iowa State College, ending with him leaving the college and Iowa to study taxidermy as an apprentice at naturalist Henry Augustus Ward's Natural Science Establishment in Rochester, New York. Between 1876 and 1879, with Ward's support, Hornaday embarked on an extensive field trip to collect specimens in India, Sri Lanka (then called Ceylon), Borneo, and on the Malay Peninsula, a period recounted in his book *Two Years in the Jungle* (1885). Now a respected taxidermist, in 1882 Hornaday was appointed chief taxidermist of the Smithsonian National Museum in Washington and later co-founded and served as the first director of the National Zoo (which also became a part of the Smithsonian Institution) in the same city. Hornaday's longest tenure was as the director of the New York Zoological Park (today the Bronx Zoo), founded in 1899, a position he held for 30 years. Hornaday was an advocate of wildlife conservation and was involved in the national campaign to save the American bison from extinction (see W. T. Hornaday, Vol. 5).[1]

Reprinted with minimal changes and a more specific title ("Wild Animal Criminals and Crime") in Hornaday's *The Minds and Manners of Wild Animals* (1922), the following article first appeared in the prominent Progressive-Era illustrated magazine *McClure's*, a monthly that marked the rise of a new form of investigative journalism – muckraking – bent on dragging illicit business practices and instances of economic and political corruption into the public spotlight.[2] Hornaday's ideas about animal "criminality" were neither particularly idiosyncratic nor without historical precedent. In "Bugs and Beasts Before the Law," an essay first published in the August 1884 issue of the *Atlantic Monthly* and later in revised form in *The Criminal Prosecution and Capital Punishment of Animals* (1906), Hornaday's contemporary Edward Payson Evans collected a significant amount of material that testified to the widespread belief in animals' moral agency and legal culpability in medieval and early modern European history.[3] Drawing on his experiences as zoo director and presenting various examples of deadly violence among captive animals – including the case of the jaguar Lopez, the "Terror of the Zoo"[4] – Hornaday argues that "the criminal propensities of mankind are mirrored in the lower animals to an astonishing degree." His assertion that such behaviour was exacerbated by "artificial conditions of captivity" apparently did not lead him to reflect upon his own moral complicity.

Notes

1 Phillip Drennon Thomas, "Hornaday, William Temple," in *American National Biography*, ed. John A. Garraty and Mark C. Carnes, vol. 11 (New York: Oxford University Press, 1999), 210–11; Gregory J. Dehler, *The Most Defiant Devil: William Temple Hornaday and His Controversial Crusade to Save American Wildlife* (Charlottesville: University of Virginia Press, 2013); Mary Anne Andrei, *Nature's Mirror: How Taxidermists Shaped America's Natural History Museums and Saved Endangered Species* (Chicago: University of Chicago Press, 2020).

2 The first issue of *McClure's* appeared in June 1893, the summer of the Panic of 1893, which threw the country into a significant economic decline that lasted until 1897 and gave the new publication a difficult start. The fortunes of the struggling magazine changed in large part due to a well-received biographical series on Napoleon written by Ida M. Tarbell, whose contributions would continue to be an important part of the magazine's future success. It was Tarbell's history of John D. Rockefeller's Standard Oil Company which turned out to be her most famous work for *McClure's*. On Tarbell, *McClure's*, and muckraking, see Bruce J. Evensen, "*McClure's Magazine*," in *Encyclopedia of American Journalism*, ed. Stephen L. Vaughn (New York: Routledge, 2008), 295–96; Harold S. Wilson, *McClure's Magazine and the Muckrakers* (Princeton: Princeton University Press, 1970); Kathleen Brady, *Ida Tarbell: Portrait of a Muckraker* (Pittsburgh: University of Pittsburgh Press, 1989).

3 By the turn of the twentieth century, the discourse of animal criminality often revolved around several western animal "outlaws" or "renegades" – wolves in particular, though not exclusively – whose cunning enabled them to elude even the experienced hunters sent after them (see D. L. Merritt, Vol. 5).

4 Ellen Velvin, "'Animals with a History': Lopez, Terror of the Zoo," *The New York Times*, December 17, 1905, 36.

11

ANIMAL CRIME AND CRIMINAL ANIMALS (1908)

William T. Hornaday

Source: William Temple Hornaday, "The Psychology of Wild Animals," *McClure's Magazine* 30, no. 4 (February 1908): 469–79.

THE PSYCHOLOGY OF WILD ANIMALS

ILLUSTRATIONS BY CARL RUNGIUS[1]

HE who enters the field of wild-animal psychology, even for a brief stroll, should pray to be delivered from the folly of seeing too much and the evil of knowing things that are not true. There is nothing more ruinous to the value of observations and deductions on the mental traits of animals than the vaulting imagination that always sees the marvelous and the boundless credulity that believes the incredible. Observations are useful only when they are squared by reason and common sense and grounded on the bed-rock of absolute truth.

While a considerable amount of serious study has been bestowed upon the mentality of the dog, the cat, the horse, and a few other domestic animals, the mental and moral moods and tenses of wild animals are but little known. It is natural for a hunter to believe that after meeting an animal twenty times in its own haunts he has acquired a fair understanding of its moral character. I know this through personal experience. But I have proved to my complete satisfaction that, no matter how well a hunter or field naturalist may learn the mental processes and traits of a free wild animal, he cannot possibly measure its intelligence nor sound the depths of its moral possibilities without making it captive and studying it in captivity.

Many human beings are "good" because they have never been under the harrow of circumstances nor sufficiently tempted to do wrong. It is only under the strain of strong temptation that human character is put through the thirty-third degree[2] and tried out. No doubt a great many of us could be provoked to join a mob for murder, or forced to steal, or tortured into insanity. It is only under the artificial conditions of captivity, with loss of freedom, exemption from the daily fear of death, abundant food without compensating labor, and with every want supplied, that the potential "cussedness" of wild creatures comes to the surface. A captive animal often reveals

DOI: 10.4324/9781003032717-12

traits never recognized in the free individual. These manifestations are of many kinds; but for the present moment we propose to consider only one.

To the lover of wild animals who no longer is troubled by the absurd question, "Do animals think?" there opens a vast and interesting field for observation and thought. Without unduly exalting the intelligence of the higher vertebrates, many thoughtful men and women have already discovered that the most interesting side of a wild animal is its mental processes as revealed in its actions and daily life.

The extent of the mental and moral parallelism between man and brute is a source of constant surprise. In a state of freedom, untrammeled by anything save the fear of death by violence, the deer or the mountain sheep works out in his own way his chosen scheme for the survival of the fittest – that is, of himself. In the wilds we see very few manifestations of the unadulterated criminal instinct. A fight between wild elk stags for the supremacy of a herd of hinds is not a manifestation of murder-lust, but of the natural instinct to multiply unmolested.

The killing of natural prey for daily food is not murder. A starving wolf on the desolate barren grounds may even kill and devour a wounded pack-mate without becoming a criminal by that act alone. True, such a manifestation of hard-heartedness and bad taste is very reprehensible; but its cause is hunger, not sheer blackness of heart. Among wild animals, the wanton killing of a member of the killer's own species constitutes murder in the first degree. Second-degree murder is unnecessary and wanton killing outside the killer's own species.

To many a wild animal there comes at times the murder-lust which under the spur of opportunity leads to genuine crime. In some of the many cases that have come under my notice, the desire to commit murder for the sake of murder has been as sharply defined as the fangs and horns of the murderers. Of the many emotions of wild animals which are revealed more sharply in captivity than in a state of nature, the crime-producing passions – jealousy, hatred, and the devilish lust for innocent blood – are most prominent. In the management of large animals in captivity, the criminal instinct is quite as great a trouble-breeder and source of anxiety as are wild-animal diseases and the constant struggle with the elements.

In many cases there is not the slightest premonitory manifestation of murderous intent on the part of an impending criminal. Indeed, with most cunning wisdom, a wild-animal murderer will often conceal his purpose until outside interference is an impossibility and the victim is entirely helpless. These manifestations of fiendish cunning and premeditation are very exasperating to those responsible for the care of animals in captivity.

In every well-regulated zoölogical park, solitary confinement is regarded as an unhappy and intolerable condition. Animals that live in herds or groups in large inclosures always exercise more, have better appetites, and are much more contented and happy than individuals that are singly confined. To visitors, a happy and contented community of deer, antelopes, bears, wolves, or birds is a source of far more pleasure than could be found in any number of solitary individuals. A small pen with a solitary animal in it at once suggests the prison-and-prisoner idea, and often arouses indignant compassion rather than pleased admiration. The peaceful community is the thing to strive for as the highest ideal attainable in an exhibition of wild animals; but mark the difficulties.

All the obstacles encountered in carrying out the community idea are created by the evil propensities of the animals themselves. Among the hoofed animals generally, every pair of horns and front hoofs is a possible storm-center. No keeper knows whether the members of his herd of deer will live together in peace and contentment until to-morrow, or whether, on any autumn or winter night, a buck will suddenly develop in his antlered head the thought that it is a good time to "kill something." In the pairing season we expect trouble, and the danger signal is always up. In October a male elk may become ever so savage, and finally develop into a raging demon, dangerous to man and beast; but when he manifests his temper *openly* and in the broad light of day, we feel that he is treating fairly both his herd-mates and his keepers. If he gives fair warning to the world about him, we must not class him as a mean criminal, no matter what he may do later on. It is our duty to corral him according to the violence of his rage. If we separate him from the herd, and he tears a fence in pieces and kills his rival, that is honest, open warfare, not foul murder. But take the following case.

In October, 1905, the New York Zoölogical Park received from the state of Washington a young mule-deer buck and two does. Being conspicuous members of the worst species of "difficult" deer to keep alive at Atlantic tidewater, and being also very thin and weak, it required the combined efforts of several persons to keep them alive. For six months they staggered about their corral, more dead than alive; but at last they began slowly to improve. The oldest doe gave birth to two fawns which actually survived. But, even when the mating season began, the buck continued to be languid and blasé. At no time did he exhibit signs of temper, or even suspicious vigor.

THE CRIME OF THE BIG JAGUAR LOPEZ

"One of the most cunning crimes I have ever seen among wild animals. . . . in an instant, Lopez rushed upon her, seized her whole neck in his powerful jaws, and crushed her cervical vertebrae by his awful bite"

In the middle of the night of November 6, 1906, without the slightest warning, he decided to commit a murder; and the mother of the two nursing fawns was selected as the victim! Being weak from the rearing of her offspring, she was at his mercy. He gored her most savagely about twenty times, and killed her.

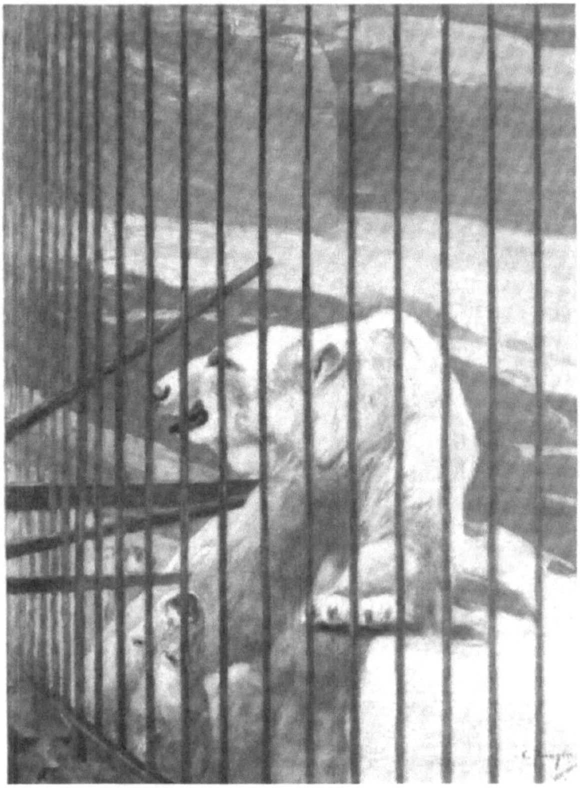

A POLAR BEAR TRAGEDY

"When she first fell, the slope of the floor brought her near the bars, which gave us a chance to fight for her. . . . but each time we beat him off, he flung himself upon her anew and crushed her down upon the snow"

That was deliberate, fiendish, and cowardly murder. The killing of any female animal by the male consort is murder; but there are circumstances wherein the plea of temporary insanity is an admissible defense. In the autumn male members of the deer family often become temporarily insane and irresponsible, and should be judged accordingly.

THE KILLING OF THE BUFFALO BULL APACHE

In the spring of 1904 apache was added to that portion of the herd in the New York Zoölogical Park of which Black Beauty was the leader. The new-comer immediately attacked Black Beauty's supremacy, and was victorious. Black Beauty awaited his opportunity, and one day, when he discovered apache lying dozing, gored him to death

Such distressing cases as the above are so common that whenever I go deer-hunting and kill a lusty buck, the thought occurs to me, "Another undeveloped murderer, perhaps!"

The most exasperating thing about these corral murders is the cunning treachery of the murderers. Take this as a typical case: For three years a dainty little male Osceola deer[3] from Florida was as gentle as a fawn and as harmless as a dove. But one crisp morning Keeper Quinn, to whom every doe in his charge is as dear as a foster-daughter, was horrified at finding blood on the absurd little antlers of the Osceola pet. One of the females lay dead in the dark corner where she had been murdered during the night; and this with another and older buck in the same corral which might fairly have been regarded as an offensive rival.

The desire to murder for the sake of killing is born in some carnivorous animals, and by others it is achieved. Among the largest and finest of the felines, the lions and tigers, midnight murders are very rare. Individual dislike is shown boldly and openly, and we are given a fair chance to prevent fatalities. Among the lions, tigers, leopards, jaguars, and pumas of the New York Zoölogical Park, twenty-eight in number, there has been but one murder. That was the crime of Lopez, the big jaguar, who deserved instant death as punishment. It was one of the most cunning crimes I have ever seen among wild animals, and is now historic.

For a year Lopez pretended ostentatiously to be a good-natured animal. Twenty times at least he acted the part of a playful pet, inviting me to reach in and stroke him. At last we decided to give him a cage-mate, and a fine adult female jaguar was purchased. The animals actually tried to caress each other through the bars, and the big male completely deceived us all.

At the end of two days it was considered safe to permit the female jaguar to enter the cage of Lopez. She was just as much deceived as we were. An animal that is afraid always leaves its traveling-cage slowly and unwillingly, or not at all. When the two sets of doors were opened, the female joyously walked into the cage of her treacherous admirer. In an instant, Lopez rushed upon her, seized her whole neck in his powerful jaws, and crushed her cervical vertebrae by his awful bite. We beat him over the head; we spiked him; we even tried to brain him: but he held her, as a bull-dog would hold a kitten, until she was dead. He had determined to murder her, but had cunningly concealed his purpose until his victim was fully in his power.

Bears usually fight "on the square," openly and aboveboard, rarely committing foul murder. If one bear hates another, he attacks at the very first opportunity; he does not cunningly wait to catch the offender at a disadvantage, when he is beyond the possibility of rescue. Bears frequently kill one another, and often maul their keepers, but not by the sneaking methods of the human assassin who stabs in the dark and runs away. I do not count the bear as a common murderer, even though, at rare intervals, he kills a cage-mate smaller and weaker than himself. One killing of that kind, done by Cinnamon Jim[4] to a small black bear that had annoyed him beyond all endurance, was inflicted as a legitimate punishment, and was so recorded. The attack of two large bears, a Syrian and a sloth bear, upon a small Japanese black bear, in which the big pair deliberately attempted to disembowel the small victim, biting him only in the abdomen, always has been a puzzle to me. I cannot fathom the idea which possessed those two ursine minds; but I have no doubt that some of the many book-making men who read the minds of animals as if they were open books could tell me all about it.

On the ice-pack in front of his stone hut at the north end of the Franz Josef Archipelago Nansen[5] saw an occurrence that was foul murder. A large male polar bear feeding upon a walrus was approached across the ice-pack by two small polar-bear cubs. The gorging male immediately stopped feeding and rushed toward the small intruders. They turned and fled wildly; but the villain pursued them far out upon the ice. He overtook them, killed both, and then serenely returned to his feast.

In February, 1907, a bear tragedy occurred in the Zoölogical Park which was a close parallel of the Lopez murder. It was a case in which my only crumb of satisfaction was in my ability to say, "I told you so," – than which no triumph can be more barren.

For seven years there lived together in the great polar bears' den of the Zoölogical Park two full-grown, very large and fine polar bears. They came from Nova Zembla,[6] and both were males. Their rough-and-tumble wrestling, both in the swimming-pool and out of it, was a sight of almost perennial interest; and while their biting and boxing was of the roughest character and frequently drew blood, they never got angry and never had a real fight.

In the autumn of 1906 one of the animals sickened and died, and the impression prevailed afterward that the survivor was lonesome. The desirability of introducing a female companion was spoken of; but I was afraid to try the experiment.

Finally, however, Mr. Karl Hagenbeck,[7] who has handled about forty polar bears to my one, wrote to us offering a fine female polar as a mate to the survivor. She was conceded to be one third smaller than the big male, but was fully adult and ready to breed. Without loss of time I answered, declining to make the purchase, on the ground that our male bear would kill the female. It was my belief that even if he did not at once deliberately murder her, he would wear her out by his rough play.

Mr. Hagenbeck replied with the assurance that, in his opinion, all would be well; that, instead of a tragedy taking place, the male would be delighted with a female companion, and that the pair would breed. As convincing proof of the sincerity of his views, Mr. Hagenbeck offered to lose half the purchase price of the female bear in the event of my worst fears being realized.

I asked the opinion of our head keeper of bears, and after due reflection he said:

> "Why, no; I don't believe he'd kill her. He's not a *bad* bear at all. I think we could work it so that there would be no great trouble."

Mr. Hagenbeck's son also felt sure there would be no tragedy.

Quite against my own judgment of polar-bear character and in deference to the expert opinion arrayed against mine, I finally yielded. The female bear was purchased, and on her arrival she was placed for three weeks in the large shifting-cage which connects with the eastern side of the great polar bears' den.

The two animals seemed glad to see each other. At once they fraternized through the bars, licked each other's nose, and ate their meals side by side. At night the male always slept as near as possible to his new companion. There was not a sign of ill temper; but, for all that, my doubts were ever present.

At last, after three full weeks of close acquaintance, it was agreed that there was nothing to be gained by longer delay in admitting the female to the large den. But we made preparations for trouble. The door of the sleeping-den was oiled and overhauled and put in thorough working order, so that if the female should dash into it for safety, a keeper could instantly slide the barrier and shut her in. We provided pike-poles, long iron bars, lariats, meat, and long planks a foot wide. Heartily wishing myself a hundred miles away, I summoned all my courage and gave the order:

"Open her door a foot only, and let her put her head out. Keep him away."

The female bear had not the slightest fear or premonition of danger. Thrusting her head through the narrow opening, she looked upon the world and the open sky above, and found that it was good. She struggled to force the door open wider; and the male stood back, waiting.

"Let her go!"

Forcing the door back with her own eager strength, she joyously dropped the intervening eighteen inches to the floor of the den, and was free. The next second the male flung his great length upon her, and the tragedy was on.

I would not for hundreds of dollars – had I the option – see such a thing again. A hundred times in the awful half-hour that followed I bitterly regretted my folly in acting contrary to my own carefully drawn conclusions regarding the temper, the strength, and the mental processes of that male bear.

He never left her for ten seconds, save when, at five or six different times, we beat him off by literally ramming him away. When she first fell, the slope of the floor brought her near the bars, which gave us a chance to fight for her. We beat him over the head; we drove big steel spikes into him; we rammed him with planks, not caring how many ribs we broke. But each time we beat him off, and the poor harried female rose to retreat, he flung himself upon her anew and crushed her down upon the snow.

Gallantly the female fought for her life, with six wild men to help her. After a long battle – it seemed like hours, but I suppose it was between twenty and thirty minutes – the male bear recognized the fact that so long as the female lay near the bars his punishment would continue and the end would be postponed. Forthwith he seized his victim and dragged her down to the ice that covered the swimming-pool in the center of the den, beyond our reach. The floor of the den was so slippery from ice and snow that it was unsafe for any of the men to enter and try to approach the now furious animal within striking distance.

Very quickly some choice pieces of fresh meat were thrown within six feet of the bears, in the hope that the male would be tempted away from his victim. In vain! Then, with all possible haste, Keeper Mulvehill coiled a lasso, entered the den, and with the first throw landed the noose neatly around the neck of the male bear. In a second it was jerked taut, the end passed through the bars, and ten eager arms dragged the big bear close to the bars. Another lariat was put on him to guard against breakages, and no bear ever missed being choked to death by a narrower margin than did that one. The morsel of revenge in it was sweet. While he was held thus, two men went in and attached a rope to the now dying female, and she was quickly dragged into the shifting-cage.

But the rescue came too late. At the last moment, on the ice, the canine teeth of the big bear had severed the jugular vein of the female, and in five minutes after her rescue she was dead.

It is my belief that at first the male did not intend to murder the female. I think his first impulse was to play with her, as he had always done with the male comrade of his own size. But the *joy of combat* seized him, and after that his only purpose was to kill. My verdict is, not premeditated murder, but murder in the second degree.

In the order of carnivorous animals, I think the worst criminals are found in the marten family (*Mustelidae*); and if there is a greater villain than the mink, I have yet to find him out. The mink is a midnight assassin who loves slaughter for the joy of murder. The wolverine, marten, mink, and weasel all are courageous, savage, and merciless. To the wolverine Western trappers accord the evil distinction of being a veritable prince of darkness on four legs. To them he is the arch-fiend beyond which animal cunning and depravity cannot go. Excepting the profane history of the pickings and stealings of this "mountain devil" as recorded by suffering trappers, we know little of it; but if its instincts are not supremely murderous, its reputation is no index of its character.

The mink, however, is a creature that we know and fear. Along the rocky shores of the Bronx River, even in the Zoölogical Park, it perversely persists. In spite of traps, guns, and poison, and the killing of from three to five annually in our park, *Putorius vison* will not down. With us, the only creatures that practise wholesale and unnecessary murder are minks and dogs. The former kill our birds, and during one awful period when a certain fence was being rebuilt the latter destroyed several deer. A mink once visited an open-air yard containing twenty-two pinioned laughing-gulls, and during that *noche trieste*[8] killed all of the ill-fated birds. But it did not devour even one, and it sucked the blood of only two or three.

On another tragic occasion a mink slaughtered an entire flock of fifteen gulls; but his joy of killing was short-lived, for he was quickly caught and clubbed to death. A miserable little weasel killed three fine brant-geese purely for the love of murder, and then he departed this life by the powder-and-lead route.

All the year round buffalo bulls are much given to fighting, and for one bull to injure or kill another is an occurrence all too common. Even in the great 27-thousand-acre preserve of the Corbin Blue Mountain Forest Association[9] fatal fights sometimes occur. But it was left to a large bull named Black Beauty,[10] in our Zoölogical Park herd, to reveal the disagreeable fact that under certain circumstances a buffalo may become a cunning and deliberate assassin.

In the spring of 1904 a new buffalo bull, named Apache, was added to the portion of our herd which up to that time had been dominated by Black Beauty. We expected the usual head-to-head battle for supremacy, succeeded by a period of peace and quiet. It is the law of the herd that after every contest for supremacy the vanquished bull shall accept the situation philosophically and thereafter keep his place.

At the end of a half-hour of fierce struggle, head to head, Black Beauty was overpowered by Apache and fled from him into the open range. To emphasize his victory, Apache followed him around and around at a quiet walk for several hours; but the beaten bull always kept a margin of safety of about four hundred feet between himself and the master of the herd. Convinced that Black Beauty would no longer dispute his supremacy, Apache at last pronounced for peace and thought no more about the late unpleasantness. His rival seemed to accept the situation, and rejoined the herd on the subdued status of ex-president.

For several days nothing occurred; but all the while Black Beauty was biding his time and watching his opportunity. At last it came. As Apache lay dozing and ruminating on a sunny slope, his beaten rival quietly drifted around his resting-place, stealthily secured a good position, and, without a second's warning, plunged his sharp horns deep into the lungs of the reclining bull. With the mad energy of pent-up and superheated fury, the assassin delivered stab after stab into the unprotected side of the helpless victim, and before Apache could gain his feet he had been gored many times. He lived only a few minutes.

It was foul murder, fully premeditated; and, had Black Beauty been my personal property, he would have been executed for the crime, without any objections or motions or appeals or far-fetched certificates of unreasonable doubt.

During the past ten years a number of persons have been foully murdered by animals they had fed and protected. The latest and one of the most deplorable of these tragedies occurred late in 1906, near Montclair, New Jersey. Mr. Herbert Bradley was the victim. While walking through his deer park, he was wantonly attacked by a white-tailed buck and murdered on the spot. At Helena, Montana, about five years ago, a strong man armed with a pitchfork was killed by a bull elk. There have been several other fatalities from elk.

The greater number of such crimes as the above have been committed by members of the deer family (deer, elk, moose, and caribou). The hollow-horned ruminants seem to be different. I believe that toward their keepers the bisons, buffaloes, and wild cattle entertain a certain sense of neutrality that in members of the deer family is, about half the time, totally absent. But there are exceptions; and a very sad and also notable case was the foul murder of Dick Rock,[11] in 1903 or thereabout.

Dick Rock was a stalwart ranchman in the prime of life, who possessed a great fondness for big-game animals. He lived near Henry Lake, Idaho, not far from the western boundary of Yellowstone Park. He liked to rope elk and moose in winter, and haul them on sleds to his ranch; to catch mountain goats and mule-deer for exhibition; and to breed buffaloes. His finest bull buffalo, named Indian, was one of his favorites, and was broken to ride. Scores of times Rock rode him around the corral, barebacked and without bridle or halter. Rock felt that he could confidently trust the animal, and he never dreamed of guarding himself against a possible evil day.

But one day the blood-lust seized the buffalo and he decided to assassinate his best friend. The next time Dick Rock entered the corral, closing the gate and fastening it securely, – thus shutting himself in, – the big bull attacked him so suddenly and so fiercely that there was not a moment for either escape or rescue. We can easily estimate the suddenness of the attack by the fact that alert and active Dick Rock had not time even to climb upon the fence of the corral, whereby his life would have been saved. With a mighty upward thrust, the treacherous bull drove one of his horns deeply into his master's body, and impaled him so completely and so securely that the man hung there and died there! As a crowning horror, the bull was unable to dislodge his victim, and the body of the ranchman was carried about the corral on the horns of his assassin until the horrified wife went a mile and a half and summoned a neighbor, who brought a rifle and executed the murderer on the spot.

Such sudden onslaughts as this make it unsafe to trust implicitly, and without recourse, to the good temper of any animal having dangerous horns.

If bird-lovers knew the prevalence of the murder instinct among the feathered folk, no doubt they would be greatly shocked. Many an innocent-looking bird is really a natural villain without opportunity to indulge in crime. It is in captivity that the inherent wickedness of wild creatures comes to the surface and becomes visible. In the open, the weak ones learn to avoid danger and to escape when threatened; but, with twenty birds in one large cage, escape is not always possible. A "happy family" of a dozen or twenty different species often harbors a criminal in its midst; and when the criminal cunningly waits until all possibilities of rescue are eliminated, an assassination is the result.

Here is a partial list of the crimes in our bird collection during 1902:

> A green jay killed a blue jay. A jay-thrush and several other small birds were killed by laughing-thrushes – which simply love to do murder! A nightingale was killed by a catbird and two mocking-birds. Two snake-birds killed a third one – all of them thoroughly depraved villains. Three gulls murdered another; a brown pelican was killed by trumpeter-swans; and a Canada goose was killed by a gull. All these victims were birds in good health.

There are a few tribes of human savages who are so far down in the moral scale that they kill their chronic invalids and their semi-helpless old men and women. It will be remembered that in our own country, Shack Nasty Jim,[12] a Modoc Indian, tomahawked his own mother because she was old and could not easily keep up with the band on its retreat from the lava-beds.

It is deplorable, but nevertheless true, that in large mixed companies of birds, say where forty or fifty live together, it is a common thing for a sick bird to be set upon and killed, unless rescued and taken away by the keepers. In crimes of this class birds often murder their own kind, but they are quite as ready to kill members of other species. In 1902 a sick brant-goose was killed by its mates; and so were a red-tailed hawk, two sarus cranes, two black vultures, a road-runner, and a great horned owl. An aged and sickly wood-ibis[13] was killed by a whooping-crane; a night-heron killed its mate; and a gull was murdered by the other members of the flock.

Strange as it may seem, among reptiles there is far less of real first-degree murder than among mammals and birds. Twenty rattlesnakes may be crowded together in one cage without a family jar. Even among cobras, perhaps the most irritable and pugnacious of all serpents, I think one snake never murders another, though they sometimes quarrel and also try to swallow one another. The big pythons and anacondas seem to know that good temper and peace promote longevity, and they almost never attack one another. And yet, a twenty-foot regal python with a bad heart – like Nansen's polar bear – could easily constrict and kill any snake of smaller size.

At this moment I do not recall one instance of wanton murder among serpents. It is well known that some snakes devour other snakes; but that is not crime. The record of the crocodilians is not so clear. It is a common thing for the large alligators and crocodiles in our reptile-house to battle for supremacy, and in these contests several fatalities have occurred. Some of these occurrences are not of the criminal sort; but when a twelve-foot alligator attacks and kills a six-foot individual, entirely out of his class and far too small to fight with him, it is murder.

Among sea creatures, the clearly defined criminal instinct, as exhibited aside from the never-ending struggle for existence and the quest of food, is rarely observed – probably because opportunities are so few. The sanguinary exploits of the grampus,[14] or whale-killer, among whales small enough to be killed and eaten, are the onslaughts of a marine glutton in quest of food.

Among the fishes there is one murderer whose evil reputation is well deserved. The common swordfish of the Atlantic, forty miles or so off Block Island or Montauk Point, is not only the most fearless of all fishes, but is also the most dangerous. His fierce attacks upon the boats of men who have harpooned him and seek to kill him are well known, and his unparalleled courage fairly challenges our wonder and admiration. But, unfortunately, the record of the swordfish is stained with crime. When the spirit of murder prompts him to commit a crime in sheer wantonness, he will attack a whale, stab the unfortunate monster again and again, and pursue it until it is dead. This is prompted solely by brutality and murder-lust, for the swordfish feeds upon fish, and never attempts to eat any portion of a whale.

Once, in the Zoölogical Park, I felt compelled to execute an animal as a measure of general safety against its criminal intent. It seemed deliberately to have resolved to do murder. A Japanese wild boar of large size and fierce disposition was kept in a corral inclosed by a fence of strong steel wire. At first the animal was just reasonably bad, and his efforts to do mischief were directed chiefly against his keepers. As time went on, however, and the supply of good food without compensating labor continued, that boar's moral character broke down completely. Like many human beings, he could not stand prosperity. But his case developed in an acute form; he became an anarchist; and his murderous mind was possessed by a desire to burst out of prescribed bounds, upset the order of things, and slay for the love of slaughter.

As he grew more and more dangerous, I watched him closely; and finally it became evident that he had resolved to break out, or die in the attempt. With a long pike-pole from the bear-dens, I stood at the corner he selected for the final breach in the steel wire, and sought to punish him until he would desist. But opposition to his will only angered him the more. With his evil little eyes fixed on me in murderous fury and the foam churning from his snapping jaws, he charged the fence again and again. Had he broken out, a crazy Malay running amuck would have been a peaceful citizen compared with him. If he should break out among the visitors, into a crowd of helpless women and children!

There was but one thing to be done. I sent a messenger at speed to my office, received from him one of my heavy rifles, and a 45-caliber bullet through the brain of the raging boar quickly ended all danger from him. With his death, I think every man in the park experienced a feeling of profound relief.

I believe it can be proved that wild animals in a perfect state of nature are not nearly so much given to wanton murder of their own kind as are some races of men. The infrequency of animal murders cannot be due wholly to the many possibilities for the intended victim to escape, nor to the difficulty of killing. In every species murders are easily possible; but it is wholly against the laws of nature for free wild beasts to kill one another in wantonness. It is left to the lower races of men to commit murders without cause, and to devour one another. The family crimes and cruelties of certain savages completely eclipse in blackness and in number the doings of the more respectable wild beasts.

In wild animals and in men, crime is an index to character. The finest species of animals and the noblest races of men are alike distinguished by their abhorrence of the shedding of innocent blood. The lion, the elephant, the wild horse, the grizzly bear, the orang-utan, the eagle, and the whooping-crane are singularly free from the criminal instinct. On the other hand, even to-day Africa is full of black or half-black tribes whose members are actually fond of practising cruelty and murder. There is to-day in the Dark Continent many a "king" beside whom a hungry lion or a grizzly bear is a noble citizen.

The vices and virtues and the criminal propensities of mankind are mirrored in the lower animals to an astonishing degree. Some of us are entitled to be classed as high as the best of the "lower" animals, but a great many creatures in human form are morally lower than the respectable four-footed beasts.

Editorial Endnotes

1 *Carl Rungius*: Carl Clemens Moritz Rungius (1869–1959), a German-born American painter and one of the foremost wildlife artists of the early twentieth century particularly known for his paintings of western American animals.

2 *put through the thirty-third degree*: a reference to the Rites of Freemasonry, specifically the Ancient and Accepted Scottish Rite of Freemasonry, where the thirty-third degree is the highest degree that can be conferred upon an individual and is reserved for members who have been of exceptional service.

3 *Osceola deer*: Florida coastal whitetail (*Odocoileus virginianus osceola*), a subspecies of white-tailed deer.

4 *Cinnamon Jim*: likely a nickname for the cinnamon bear, a subspecies of American black bear whose name is derived from its reddish-brown fur.

5 *Nansen*: Fridtjof Nansen (1861–1930), a Norwegian scientist, explorer, and humanitarian who explored Franz Josef Land, an archipelago in the Arctic Ocean named after Habsburg monarch Franz Josef I (emperor of Austria and king of Hungary) during an 1893–96 expedition.

6 *Nova Zembla*: Novaya Zemlya ("New Land"), a Russian archipelago in the Arctic Ocean.

7 *Karl Hagenbeck*: Karl Hagenbeck (1844–1913) and his younger brother Wilhelm (1850–1910) were German traders of wild animals, whom they sold to European and American zoos and circuses, including that of showman Phineas Taylor Barnum. Hagenbeck owned a zoo in Hamburg (*Hagenbecks Tierpark*), which exhibited both "exotic" animals and "primitive" humans. His "collection" was presented at the World's Columbian Exposition (Chicago, 1893) and the Louisiana Purchase Exposition (St. Louis, 1904). Wilhelm is credited by his brother with developing new animal training techniques, which included the training of polar bears.

8 noche triste: Spanish for "sad night," specifically in reference to an event during the Spanish *conquista* in which Hernán Cortés (1485–1547) and his men were expelled from the Aztec capital of Tenochtitlán on the night of June 30, 1520.

9 *Corbin Blue Mountain Forest Association*: also known as "Corbin Park," a private wildlife preserve in New Hampshire created by entrepreneur Austin Corbin (1827–1896).

10 *Black Beauty*: a reference to English novelist Anna Sewell's (1820–1878) widely read novel *Black Beauty* (1877), a fictional autobiography of an eponymous horse that conveyed a clear animal welfare message.

11 *Dick Rock*: also known as "Rocky Mountain Dick," Richard W. Rock (ca. 1846–1902) was a hunter, trapper, and former scout with the U.S. Army in campaigns against western Indians. Rock's isolated ranch near Yellowstone Park was populated by bison, bears, elk, and other creatures he had caught personally and some of whom he would sell to zoos and parks.

12 *Shack Nasty Jim*: Shacknasty Jim, or Shkeitko (ca. 1851–1881), was a Modoc Indian warrior. He fought in the Modoc War (1872–1873) in Oregon and California and, following the Modoc's defeat, led his band to the Quapaw Indian Agency in modern-day Oklahoma.

13 *wood-ibis*: today more commonly (and correctly) known as a "wood stork," a species of bird found in South and Central America and the Caribbean, with smaller populations in parts of the southern U.S.

14 *grampus*: an older name for orcas, highly social apex predators known for cooperatively hunting various aquatic mammals, including whales.

Editorial Headnote

12. Margaret Floy Washburn, *The Animal Mind: A Text-Book of Comparative Psychology*. New York: The Macmillan Company, 1908.

An influential figure in comparative psychology through the first half of the twentieth century before her perspectives on animal consciousness began to clash with the rising tide of behaviorism, Margaret Floy Washburn (1871–1939) was drawn to the emerging field of experimental psychology early on during her studies at Vassar College. Her desire to pursue her graduate studies under psychologist James McKeen Cattell at Columbia University was thwarted by the fact that the university did not accept women as graduate students, resulting in her move to Cornell University in 1892, where she received her doctorate in 1894. After working at Wells College (Aurora, NY), Cornell, and the University of Cincinnati, she returned to Vassar in 1903. Throughout her academic career, Washburn served as president of the American Psychological Association and was appointed to the prestigious National Academy of Science.

The Animal Mind was Washburn's most influential work and remained the standard textbook of comparative psychology for several decades following its 1908 publication. The book is a comprehensive engagement with the comparative psychological research of the time, including numerous German- and French-language works. Its chapters contain references to a broad diversity of organisms – around 100 different species – ranging from crayfish to cows and from earthworms to elephants. The success of Washburn's book resulted in a second edition in 1917 and in significantly expanded third (1926) and fourth (1936) editions that incorporated a significant amount of new research.[1]

The first two chapters of *The Animal Mind*, from which the following excerpts are taken, grapple with the difficulties and pitfalls that present themselves in the study of animal minds as a project in which the broader epistemological problem of other minds is compounded by additional layers or degrees of species difference. Washburn's position here is unambiguous – and arguably not far off from more recent ideas about an inevitable human-centeredness (grounded in our sensorimotor and perceptual-experiential specificity as human beings) that forms the basis of all engagements with nonhuman minds: "all psychic interpretation of animal behavior must be on the analogy of human experience."[2]

Notes

1 See Deborah Johnson, "Washburn, Margaret Floy," in *American National Biography*, ed. John A. Garraty and Mark C. Carnes, vol. 22 (New York: Oxford University Press, 1999), 747–49; Wayne Viney and Laura Burlingame-Lee, "Margaret Floy Washburn: A Quest for the Harmonies in the Context of a Rigorous Scientific Framework,"

in *Portraits of Pioneers in Psychology*, ed. Gregory Adams Kimble and Michael Wertheimer, vol. 5 (Washington: American Psychological Association, 2003), 73–88; David A. Washburn, "The Animal Mind at 100," *The Psychological Record* 60, no. 2 (2010): 369–76.

2 Emphasis removed. On inevitable anthropocentrism, see, for example, Val Plumwood, *Environmental Culture: The Ecological Crisis of Reason* (London: Routledge, 2002), chapter 6.

12

METHODS AND CHALLENGES IN STUDYING ANIMAL MINDS (1908)[1]

Margaret F. Washburn

Source: Margaret Floy Washburn, *The Animal Mind: A Text-Book of Comparative Psychology* (New York: The Macmillan Company, 1908), 1–5, 7–13, 24–36.

THE DIFFICULTIES AND METHODS OF COMPARATIVE PSYCHOLOGY

§ 1. *Difficulties*

THAT the mind of each human being forms a region inaccessible to all save its possessor, is one of the commonplaces of reflection. His neighbor's knowledge of each person's mind must always be indirect, a matter of inference. How wide of the truth this inference may be, even under the most favorable circumstances, is also an affair of everyday experience: each of us can judge his fellow-men only on the basis of his own thoughts and feelings in similar circumstances, and the individual peculiarities of different members of the human species are of necessity very imperfectly comprehended by others. The science of human psychology has to reckon with this unbridgable gap between minds as its chief difficulty. The psychologist may look into his own mind and study its workings with impartial insight, yet he can never be sure that the laws which he derives from such a study are not distorted by some personal twist or bias. For example, it has been suggested that the philosopher Hume[2] was influenced by his tendency toward a visual type of imagination in his discussion of the nature of ideas, which to him were evidently visual images. As is well known, the experimental method in psychology has aimed to minimize the danger of confusing individual peculiarities with general mental laws. In a psychological experiment, an unbiassed observer is asked to study his own experience under certain definite conditions, and to put it into words so that the experimenter may know what the contents of another mind are like in the circumstances. Thus language is the essential apparatus in experimental psychology; language with all its defects, its ambiguity, its substitution of crystallized concepts for the protean flux of actually lived experience, its lack of terms to express those parts of experience which are of small practical importance in everyday life, but which may be of the highest importance to mental science.

DOI: 10.4324/9781003032717-13

Outside of the psychological laboratory language is not always the best guide to the contents of other minds, because it is not always the expression of a genuine wish to communicate thought. "Actions speak louder than words," the proverb says; but when words are backed by good faith they furnish by far the safest indication of the thought of others. Whether, however, our inferences are made on the basis of words or of actions, they are all necessarily made on the hypothesis that human minds are built on the same pattern, that what a given word or action would mean for my mind, this it means also for my neighbor's mind.

If this hypothesis be uncertain when applied to our fellow human beings, it fails us utterly when we turn to the lower animals. If my neighbor's mind is a mystery to me, how great is the mystery which looks out of the eyes of a dog, and how insoluble the problem presented by the mind of an invertebrate animal, an ant or a spider! We know that such minds must differ from ours not only in certain individual peculiarities, but in ways at whose nature we can only guess. The nervous systems of many animals vary widely from our own. We have, perhaps, too little knowledge about the functions of our own to conjecture with any certainty what difference this must make in the conscious life of such animals; but when we find sense organs, such as the compound eyes of insects or crustaceans, constructed on a plan wholly diverse from that of ours; when we find organs apparently sensory in function, but so unlike our own that we cannot tell what purpose they serve, – we are baffled in our attempt to construct the mental life of the animals possessing them, for lack of power to supply the sensation elements of that life. "It is not," said Locke, "in the power of the most exalted wit or enlarged understanding, by any quickness or variety of thought, to invent or frame one new simple idea in the mind" [...] we cannot imagine a color or a sound or a smell that we have never experienced; how much less the sensations of a sense radically different from any that we possess! Again, a bodily structure entirely unlike our own must create a background of organic sensation which renders the whole mental life of an animal foreign and unfamiliar to us. We speak, for example, of an "angry" wasp. Anger, in our own experience, is largely composed of sensations of quickened heart beat, of altered breathing, of muscular tension, of increased blood pressure in the head and face. The circulation of a wasp is fundamentally different from that of any vertebrate. The wasp does not breathe through lungs, it wears its skeleton on the outside, and it has the muscles attached to the inside of the skeleton. What is anger like in the wasp's consciousness? We can form no adequate idea of it.

To this fundamental difficulty of the dissimilarity between animal minds and ours is added, of course, the obstacle that animals have no language in which to describe their experience to us. Where this unlikeness is greatest, as in the case of invertebrate animals, language would be of little use, since we could not interpret it from our experience; but the higher vertebrates could give us much insight into their minds if they could only speak. We are, however, restricted to the inferences we can draw from movements and sounds that are made for the most part without the intention of communicating anything to us. One happy consequence of this fact, which to a slight extent balances its disadvantages, is that we have not to contend with self-consciousness and posing, which often invalidate human reports of introspection.

From these general considerations we can understand something of the special difficulties that beset the path of the comparative psychologist, who desires to know the contents of minds below the human level. Knowledge regarding the animal mind, like knowledge of human minds other than our own, must come by way of inference from behavior. Two fundamental questions then confront the comparative psychologist. First, by what method shall he find out how an animal behaves? Second, how shall he interpret the conscious aspect of that behavior?

§ 2. *Methods of Obtaining Facts: The Method of Anecdote*

The reading of such a book as Romanes's "Animal Intelligence," or of the letters about animal behavior in the London *Spectator,* will reveal one method of gathering information about what animals do. This has been termed the Method of Anecdote. It consists essentially in taking the report of another person regarding the action of an animal, observed most commonly by accident, and attracting attention because of its unusual character. In certain cases the observer while engaged in some other pursuit happens to notice the singular behavior of an animal, and at his leisure writes out an account of it. In others, the animal is a pet, in whose high intellectual powers its master takes pride. It is safe to say that this method of collecting information always labors under at least one, and frequently under several, of the following disadvantages: –

1. The observer is not scientifically trained to distinguish what he sees from what he infers.
2. He is not intimately acquainted with the habits of the species to which the animal belongs.
3. He is not acquainted with the past experience of the individual animal concerned.
4. He has a personal affection for the animal concerned, and a desire to show its superior intelligence.
5. He has the desire, common to all humanity, to tell a good story.

[. . .]

Indeed, a root of evil in the method of anecdote consists in the fact that observation in this form is imperfectly divorced from interpretation. The maker of an anecdote is seldom content with merely telling one what the animal did and leaving future investigation and the comparative study of many facts to decide what the animal's conscious experience in doing it was like. The point of the anecdote usually consists in showing that a human interpretation of the animal's behavior is possible. Here is shown the desire to tell a good story, which we mentioned among the pitfalls of the anecdotal method; the wish to report something unusual, not to get a just conception of the normal behavior of an animal. As Thorndike has forcibly put it: "Dogs get lost hundreds of times and no one ever notices it or sends an account of it to a scientific magazine. But let one find his way from Brooklyn to Yonkers and the fact immediately becomes a circulating anecdote. Thousands of cats on thousands of occasions sit helplessly yowling, and no one takes thought of it or writes to his friend, the professor; but let one cat claw at the knob of a door supposedly as a signal to be let out, and straightway this cat becomes the representative of the cat-mind in all the books".

All this is not to deny that much of the testimony to be found in Romanes's "Animal Intelligence" and Darwin's "Descent of Man" is the trustworthy report of trained observers; but it is difficult to separate the grain from the chaff, and one feels toward many of the anecdotes the attitude of scepticism produced, for example, by this tale which an Australian lady reported to the Linnæan Society. The burial of some deceased comrades was accomplished, she says, by a nest of "soldier ants" near Sydney, in the following fashion. "All fell into rank walking regularly and slowly two by two, until they arrived at the spot where lay the dead bodies. . . . Two of the ants advanced and took up the dead body of one of their comrades; then two others, and so on until all were ready to march. First walked two ants bearing a body, then two without a burden; then two others with another dead ant, and so on, until the line was extended to about forty pairs, and the procession now moved slowly onward, followed by an irregular body of about two hundred ants. Occasionally the two laden ants stopped, and laying down the dead ant, it was taken up by the two walking unburdened behind them, and thus, by occasionally relieving each other, they arrived at a sandy spot near the sea." A separate grave was then dug for each dead ant. "Some six or seven of the ants had attempted to run off without performing their share of the task of digging; these were caught and brought back, when they were at once attacked by the body of ants and killed upon the spot. A single grave was quickly dug and they were all dropped into it." No funeral procession for them! Of this story Romanes says, "The observation seems to have been one about which there could scarcely have been a mistake". One is inclined to think it just possible that there was.

§ 3. *Methods of Obtaining Facts: The Method of Experiment*

Diametrically opposed to the Method of Anecdote and its unscientific character is the Method of Experiment. An experiment, properly conducted, always implies that the conditions are controlled, or at least known; whereas ignorance of the conditions is, as we have seen, a common feature of anecdote. The experimenter is impartial; he has no desire to bring about any particular result. The teller of an anecdote wishes to prove animal intelligence. The experimenter is willing to report the facts precisely as he observes them, and is in no haste to make them prove anything. The conduct of an experiment upon an animal will, of course, vary according to the problem to be solved. If the object is to test some innate reaction on the animal's part, such as its ordinary responses to stimulation or its instincts, one need merely place the animal under favorable conditions for observation, make sure that it is not frightened or in an abnormal state, supply the appropriate stimulus unmixed with others, and watch the result. If it is desired to study the process by which an animal learns to adapt itself to a new situation, one must, of course, make sure in addition that the situation really is new to the animal, and yet that it makes sufficient appeal to some instinctive tendency to supply a motive for the learning process.

[. . .]

Wesley Mills,[3] professor of physiology in McGill University, has studied very carefully the mental development of young animals such as cats and dogs, but is inclined to criticise the use of experiment in observing animals. The work of E. L. Thorndike, whose "Animal Intelligence" appeared in 1898, represents, perhaps, the first definite effect of the modem experimental movement in psychology upon the study of the animal mind. Thorndike's aim in this research was to place his animals (chicks, cats, and dogs) under the most rigidly controlled experimental conditions. The cats and dogs, reduced by fasting to a state of "utter hunger," were placed in boxes, with food outside, and the process whereby they learned to work the various mechanisms which let them out was carefully observed. Since the appearance of Thorndike's work the performance of experiments upon animals has played much part in the work of psychological laboratories, particularly those of Harvard, Clark, and Chicago universities. The biologists and physiologists have continued their researches by this method, so that a very large amount of experimental work is now being done in comparative psychology.

Despite the obvious advantages of experiment as a method for the study of animal behavior, it is not without its dangers. These were clearly stated by Wesley Mills in a criticism of Thorndike's "Animal Intelligence". They may be summed up by saying that there is a risk of placing the animal experimented upon under abnormal conditions in the attempt to make them definite and controllable. Did not, for example, the extreme hunger to which Thorndike's cats and dogs were reduced, while it simplified the conditions in one sense by making the strength of the motive to escape as nearly as possible equal for all the animals, complicate matters in another sense by diminishing their capacity to learn? Were the animals perhaps frightened and distracted by the unusual character of their surroundings? Thorndike thinks not; but whether or no he succeeded in averting these dangers, it is clear that they are real. It is also obvious that they are the more threatening, the higher the animal with which one has to deal. Fright, bewilderment, loneliness, are conditions more apt to be met with among the higher vertebrates than lower down in the scale, and the utmost care should be taken to make sure that animals likely to be affected by them are thoroughly trained and at home in their surroundings before the experimenter records results.

§ 4. *Methods of Obtaining Facts: The Ideal Method*

The ideal method for the study of a higher animal involves patient observation upon a specimen known from birth, watched in its ordinary behavior and environment, and occasionally experimented upon with proper control of the conditions and without frightening it or otherwise rendering it abnormal. The observer should acquaint himself with the individual peculiarities of each animal studied, for there is no doubt that striking differences in mental capacity occur among the individuals of a single species. At the same time that he obtains the confidence of each individual animal, he should be able to hold in check the tendency to humanize it and to take a personal pleasure in its achievements if it be unusually endowed. This is, to say the least, not easy. Absolute indifference to the animals studied, if not so dangerous as doting affection, is yet to be avoided.

§ 5. *Methods of Interpreting Facts*

We may now turn from the problem of discovering the facts about animal behavior to the problem of interpreting them. If an animal behaves in a certain manner, what may we conclude the consciousness accompanying its behavior to be like? As we have seen, the interpretation is often confused with the observation, especially in the making of anecdotes; but theoretically the two problems are distinct. And at the outset of our discussion of the former, we are obliged to acknowledge that *all psychic interpretation of animal behavior must be on the analogy of human experience.* We do not know the meaning of such terms as perception, pleasure, fear, anger, visual sensation, etc., except as these processes form a part of the contents of our own minds. Whether we will or no, we must be anthropomorphic in the notions we form of what takes place in the mind of an animal. Accepting this fundamental proposition, the students of animals have yet differed widely in the conclusions they have drawn from it. Some have gone to the extreme of declaring that comparative psychology is therefore impossible. Others have joyfully hastened to make animals as human as they could. Still others have occupied an intermediate position.

[. . .]

Is it possible to state briefly the special precautions that must be observed in interpreting animal behavior as accompanied by consciousness, granted that such interpretation is admissible? Jordan,[4] while holding that the existence of the animal mind may fairly be inferred under certain circumstances, holds that we are not justified in inferring the actual *quality* of mental processes in animals. For this reason he objects to the term "comparative psychology". There is no doubt that great caution should be used in regarding the quality of a human conscious process as identical with the quality of the corresponding process in the animal mind. For example, we might say with a fair degree of assurance that an animal consciously discriminates between light and darkness; that is, receives conscious impressions of different quality from the two, yet the mental impression produced by white light upon the animal may be very different from the sensation of white as we know it, and the impression produced by the absence of light very different from our sensation of black. Black and white may, for all we know, depend for their quality upon some substance existing only in the human retina.

A second precaution concerns the simplicity or complexity of the interpretation put upon animal behavior. Lloyd Morgan, in his "Introduction to Comparative

Psychology," formulated a conservative principle of interpretation which has often been quoted as "Lloyd Morgan's Canon." The principle is as follows: "In no case may we interpret an action as the outcome of the exercise of a higher psychical faculty, if it can be interpreted as the outcome of the exercise of one which stands lower in the psychological scale". In other words, when in doubt take the simpler interpretation. For example, a dog detected in a theft cowers and whines. One possible mental accompaniment of this behavior is remorse; the dog is conscious that he has fallen below a moral standard, and grieved or offended his master. A second is the anticipation of punishment; the dog has a mental representation of the consequences of his action upon former occasions, and imagining himself likely to experience them anew, is terrified at the prospect. A third possibility is that the dog's previous experience of punishment, instead of being revived in the form of definite images, makes itself effective merely in his feelings and behavior; he is uncomfortable and frightened, he knows not definitely why. It is evident that these three possibilities represent three different grades of complexity of mental process, the first being by far the highest. Lloyd Morgan's canon enjoins upon us in such a case to prefer the third alternative, provided that it will really account for the dog's behavior.

Now why should the simplest interpretation be preferred? We must not forget that the more complex ones remain in the field of possibility. Positive assertions have no place in comparative psychology. We cannot say that the simplicity of an hypothesis is sufficient warrant of its truth, for nature does not always proceed by the paths which seem to us least complicated. The fact is that Lloyd Morgan's principle serves to counterbalance our most important source of error in interpreting animal behavior. It is like tipping a boat in one direction to compensate for the fact that some one is pulling the opposite gunwale. We must interpret the animal mind humanly if we are to interpret it at all. Yet we know that it differs from the human mind, and that the difference is partly a matter of complexity. Let us therefore take the least complex interpretation that the facts of animal behavior will admit, always remembering that we may be wrong in so doing, but resting assured that we are, upon the whole, on the safer side. The social consciousness of man is very strong, and his tendency to think of other creatures, even of inanimate nature, as sharing his own thoughts and feelings, has shown itself in his past to be almost irresistible. Lloyd Morgan's canon offers the best safeguard against this natural inclination, short of abandoning all attempt to study the mental life of the lower animals.

CHAPTER II

THE EVIDENCE OF MIND

§ 6. *Inferring Mind from Behavior*

In the last chapter we saw that some recent writers upon animal behavior and its interpretation, while refusing to admit the presence of consciousness in all forms of animal life, yet hold that it can be proved to exist in certain forms. The latter, it is maintained, display certain peculiarities of behavior that may be regarded as proofs of a psychic accompaniment. Into the nature of these proofs we may now inquire.

To begin with, can it be said that when an animal makes a movement in response to a certain stimulus, there is an accompanying consciousness of the stimulus, and that when it fails to move, there is no consciousness? Is *response to stimulation* evidence of consciousness? In the case of man, we know that absence of visible response does not prove that the stimulus has not been sensed; while it is probable that some effect upon motor channels always occurs when consciousness accompanies stimulation, the effect may not be apparent to an outside observer. On the other hand, if movement in response to the impact of a physical force is evidence of consciousness, then the ball which falls under the influence of gravity and rebounds on striking the floor is conscious. Nor is the case improved if we point out that the movements which animals make in response to stimulation are not the equivalent in energy of the stimulus applied, but involve the setting free of energy stored in the animal as well. True, when a microscopic animal meets an obstacle in its swimming, and darts backward, the movement is not a mere rebound; it implies energy contributed by the animal's own body. But just so an explosion of gunpowder is not the equivalent in energy of the heat of the match, the stimulus. Similarly it is possible to think of the response made by animals to external stimuli as involving nothing more than certain physical and chemical processes identical with those existing in inanimate nature.

If we find that the movements made by an animal as a result of external stimulation regularly involve withdrawal from certain stimuli and acceptance of others, it is natural to use the term "*choice*" in describing such behavior. But if consciousness is supposed to accompany the exercise of choice in this sense, then consciousness must be assumed to accompany the behavior of atoms in chemical combinations. When hydrochloric acid is added to a solution of silver nitrate, the atoms of chlorine and those of silver find each other by an unerring "instinct" and combine into the white precipitate of silver chloride, while the hydrogen and the nitric acid similarly "choose" each other. Nor can the fact that behavior in animals is adapted to an end be used as evidence of mind; for "purposive" reactions, which contribute to the welfare of an organism, are themselves selective. The search for food, the care for the young, and the complex activities which further welfare, are made up of reactions involving "choice" between stimuli; and if the simple "choice" reaction is on a par with the behavior of chemical atoms, so far as proof of consciousness goes, then *adaptation to an end*, apparent purposiveness, is in a similar position.

Thus the mere fact that an animal reacts to stimulation, even selectively and for its own best interests, offers no evidence for the existence of mind that does not apply equally well to particles of inanimate matter. Moreover, there is some ground for holding that the reactions of the lowest animals are unconscious. This ground consists in the apparent lack of variability which characterizes such reactions. In our own case, we know that certain bodily movements, those of digestion and circulation, for example, are normally carried on without accompanying consciousness, and that in other cases where there is consciousness of the stimulus, as in the reflex knee-jerk, it occurs after the movement is initiated, so that the nervous process underlying the sensation would seem to be immaterial to the performance of the movement. These unconscious reactions in human beings are characterized by their relative uniformity, by the absence of variation in their performance. Moreover, when an action originally accompanied by consciousness is often repeated, it tends, by what is apparently one and the same process, to become *unconscious* and to become uniform. There is consequently reason for believing that when the behavior of lower animals displays perfect uniformity, consciousness is not present. On the other hand, an important reservation must be made in the use of this negative test. It is by no means easy to be sure that an animal's reactions are uniform. The more carefully the complexer ones are studied, the more are variability and difference brought to light where superficial observation had revealed a mechanical and automatic regularity. It is quite possible that even in the simple, apparently fixed response of microscopic animals to stimulation, better facilities for observation might show variations that do not now appear.

This matter of uniformity *versus* variability suggests a further step in our search for a satisfactory test of the presence of mind. Is mere *variability* in behavior, mere irregularity in response, to be taken as such a test? Not if we argue from our own experience. While that portion of our own behavior which involves consciousness shows more irregularity than the portion which does not, yet the causes of the irregularity are often clearly to be found in physiological conditions with which consciousness has nothing to do. There are days when we can think clearly and recall easily, and days when obscurities refuse to vanish and the right word refuses to come; days when we are irritable and days when we are sluggish. Yet since we can find nothing in our mental processes to account for this variability, it would be absurd to take analogous fluctuations in animal behavior as evidence of mind. So complicated a machine as an animal organism, even if it be nothing more than a machine, must show irregularities in its working.

Behavior, then, must be variable, but not merely variable, to give evidence of mind. The criterion most frequently applied to determine the presence or absence of the psychic is *a variation in behavior that shows definitely the result of previous individual experience.* "Does the organism," says Romanes, "learn to make new adjustments, or to modify old ones, in accordance with the results of its own individual experience?" (364, p. 4). Loeb[5] declares that "the fundamental process which occurs in all psychic phenomena as the elemental component" is "the activity of the associative memory, or of association," and defines associative memory as "that mechanism by which a stimulus brings about not only the effects which its nature and the specific structure of the irritable organ call for, but by which it brings about also the effects of other stimuli which formerly acted upon the organism almost or quite simultaneously with the stimulus in question."

"If an animal can be trained," he continues, "if it can learn, it possesses associative memory," and therefore mind. The psychologist finds the term "associative memory" hardly satisfactory, and objects to the confusion between mental and physical concepts which renders it possible to speak of a "mechanism" as forming an "elemental component" in "psychic phenomena," but these points may be passed over. The power to learn by individual experience is the evidence which Romanes, Morgan, and Loeb will accept as demonstrating the presence of mind in an animal.

Does the absence of proof that an animal learns by experience show that the animal is unconscious? Romanes is careful to answer this question in the negative. "Because a lowly organized animal," he says, "does *not* learn by its own individual experience, we may not therefore conclude that in performing its natural or ancestral adaptations to appropriate stimuli, consciousness, or the mind element, is wholly absent; we can only say that this element, if present, reveals no evidence of the fact". Loeb, on the other hand, writes as if absence of proof for consciousness amounted to disproof, evidently relying on the principle of parsimony, that no unnecessary assumptions should be admitted. "Our criterion," he remarks, "puts an end to the metaphysical ideas that all matter, and hence the whole animal world, possesses consciousness". If learning by experience be really a satisfactory proof of mind, then its. absence in certain animals would indeed prevent the positive assertion that all animals are conscious; but it could not abolish the possibility that they might be. Such a possibility might, however, be of no more scientific interest than any one of a million wild possibilities that science cannot spare time to disprove. But we shall find that learning by experience, taken by itself, is too indefinite a concept to be of much service, and that when defined, it is inadequate to bear the whole weight of proving consciousness in animals. Such being the case, the possibility that animals which have not been shown to learn may yet be conscious acquires the right to be reckoned with.

The first point that strikes us in examining the proposed test is that the learning by experience must not be too slow, or we can find parallels for it in the inanimate world. An animal may be said to have learned by experience if it behaves differently to a stimulus because of preceding stimuli. But it is one thing to have behavior altered by a single preceding stimulus, and another to have it altered by two hundred repetitions of a stimulus. The wood of a violin reacts differently to the vibrations of the strings after it has "experienced" them for ten years; the molecules of the wood have gradually taken on an altered arrangement. A steel rail reacts differently to the pounding of wheels after that process has been long continued; it may snap under the strain. Shall we say that the violin and the rail have learned by individual experience? If the obvious retort be made that it is only in living creatures that learning by experience should be taken as evidence of mind, let us take an example from living creatures. When a blacksmith has been practising his trade for a year, the reactions of his muscles are different from what they were at the outset. But this difference is not merely a matter of more accurate sense-discrimination, a better "placing" of attention and the like; there have been going on within the structure of his muscles changes which have increased their efficiency, and with which consciousness has had nothing to do. These changes have been extremely slow compared to the learning which does involve consciousness. In one or two lessons the apprentice learned what he was to do; but only very gradually have his muscles acquired the strength to do it as it should be done. Now among the lower animal forms we sometimes meet with learning by experience that is very slow; that requires a hundred or more repetitions of the stimulus before the new reaction is acquired. In such a case we can find analogical reasons for suspecting that a gradual change in the tissues of the body has taken place, of the sort which, like the attuning of the violin wood or the slow development of a muscle, have no conscious accompaniment.

We must then ask the question: *What kind of learning by experience never,* so far as we know, *occurs unconsciously?* Suppose a human being shut up in a room from which he can escape only by working a combination lock. As we shall see later, this is one of the methods by which the learning power of animals has been tested. The man, after prolonged investigation, hits upon the right combination and gets out. Suppose that he later finds himself again in the same predicament, and that without hesitation or fumbling he opens the lock at once, and performs the feat again and again, to show that it was not a lucky accident. But one interpretation of such behavior is possible. We know from our own experience that the man could not have worked the lock the second time he saw it, unless he consciously remembered the movements he made the first time; that is, unless he had in mind some kind of idea as a guide. Here, at least, there can have been no change in the structure of the muscles, for such changes are gradual; the change must have taken place in the most easily alterable portion of the organism, the nervous system; and further, it must have taken place in the most unstable and variable part of the nervous system, the higher cortical centres whose activity is accompanied by consciousness. In other words, we may be practically assured that consciousness accompanies learning only when the learning is so rapid as to show that the effects of previous experience are recalled in the guise of an idea or mental image of some sort. But does even the most rapid learning possible assure us of the presence of an idea in the mind of a lower animal? Where the motive, the beneficial or harmful consequence of action, is very strong, may not a single experience suffice to modify action without being revived in idea? Moreover, animals as high in the scale as dogs and cats learn to solve problems analogous to that of the combination lock so slowly that we cannot infer the presence of ideas. Are we then to conclude that these animals are unconscious, or that there is absolutely no reason for supposing them possessed of consciousness? Yerkes[6] has criticised the "learning by experience" criterion by pointing out that "no organism . . . has thus far been proved incapable of profiting by experience." It is a question rather of the rapidity and of the kind of learning involved. "The fact that the crayfish needs a hundred or more experiences for the learning of a type of reaction that the frog would learn with twenty experiences, the dog with five, say, and the human subject with perhaps a single experience, is indicative of the fundamental difficulty in the use of this sign". Nagel[7] has pointed out that Loeb, in asserting "associative memory" as the criterion of consciousness, offers no evidence for his statement. The fact is that while proof of the existence of mind can be derived from animal learning by experience only if the learning is very rapid, other evidence, equally valid on the principle of analogy, makes it *highly improbable that all animals which learn too slowly to evince the presence of ideas are therefore unconscious.* This evidence is of a *morphological* character.

§ 7. *Inferring Mind from Structure*

Both Yerkes and Lukas[8] urge that the resemblance of an animal's nervous system and sense-organs to those of human beings ought to be taken into consideration in deciding whether the animal is conscious or not. Lukas suggests that the criteria of consciousness should be grouped under three heads: morphological, including the structure of the brain and sense-organs, physiological, and teleological. Under the second rubric he maintains that "individual purposiveness" is characteristic of the movements from which consciousness may be inferred; that individual purposiveness pertains only to voluntary acts, and that voluntary acts are acts "which are preceded by the intention to perform a definite movement, hence by the idea of this movement." We have reached the same conclusion in the preceding paragraph. The third test of the presence of consciousness, the teleological test, rests on the consideration: "What significance for the organism may be possessed by the production of a conscious effect by certain stimuli?". This test, however, being of a purely *a priori* character, would seem to be distinctly less valuable than the others.

Yerkes proposes "the following six criteria in what seems to me in general the order of increasing importance. The functional signs are of greater value as a rule than the structural; and within each of the categories the particular sign is usually of more value than the general. In certain cases, however, it might be maintained that neural specialization is of greater importance than modifiability.

I. Structural Criteria.

   1. General form of organism (Organization).
   2. Nervous system (Neural organization).
   3. Specialization in the nervous system (Neural specialization).

II. Functional Criteria.

   1. General form of reaction (Discrimination).
   2. Modifiability of reaction (Docility).
   3. Variability of reaction (Initiative)".

The terms "discrimination," "docility," and "initiative" in this connection are borrowed from Royce's[9] "Outlines of Psychology".

If resemblance of nervous and sense-organ structure to the human type is to be taken along with rapid learning as coordinate evidence of consciousness, it is clear that here also we have to deal with a matter of degree. The structure of the lower animals differs increasingly from our own as we go down the scale. At what degree of difference shall we draw the line and say that the animals above it may be conscious, but that those below it cannot be? No one could possibly establish such a line. The truth of the whole matter seems to be this: *We can say neither what amount of resemblance in structure to human beings, nor what speed of learning, constitutes a definite mark distinguishing animals with minds from those without minds, unless we are prepared to assert that only animals which learn so fast that they must have memory ideas possess mind at all.* And this would conflict with the argument from structure. For example, there is no good experimental evidence that cats possess ideas, yet there is enough analogy between their nervous systems and our own to make it improbable that consciousness, so complex and highly developed in us, is in them wholly lacking. We know not where consciousness begins in the animal world. We know where it surely resides – in ourselves; we know where it exists beyond a reasonable doubt – in those animals of structure resembling ours which rapidly adapt themselves to the lessons of experience. Beyond this point, for all we know, it may exist in simpler and simpler forms until we reach the very lowest of living beings.

Editorial Endnotes

1 Washburn's book includes in-text references to a numbered bibliography. For reasons of readability, these have been silently removed in this excerpt.

2 *Hume*: David Hume (1711–1776), a prominent Scottish Enlightenment philosopher known for his work in epistemology (as a strong proponent of philosophical skepticism), moral philosophy, and other areas. Hume had a direct influence on the thought of major figures of the eighteenth and nineteenth centuries, including Adam Smith, Kant, and Charles Darwin.

3 *Wesley Mills*: Thomas Wesley Mills (1847–1915) was a Canadian physician and physiologist. His chief contribution in the area of comparative psychology consists in a monograph on *The Nature and Development of Animal Intelligence* (1898). Mills argued that animal intelligence had to be tested in an animal's natural environment rather than the kind of artificial experimental scenarios developed by people like Edward L. Thorndike.

4 *Jordan*: Hermann Jacques Jordan (1877–1943), a German-Dutch physiologist and zoologist who held a professorship of comparative psychology at the University of Utrecht from 1919 until 1943, when he was fired and forced to go into hiding with his Jewish wife.

5 *Loeb*: Jacques Loeb (1859–1924), a German-born American physiologist and biologist noted for his work in experimental embryology and neurophysiology. Loeb received a total of 78 nominations for the Nobel Prize in Physiology and Medicine.

6 *Yerkes*: Robert Mearns Yerkes (1876–1956), an American psychologist and ethologist and a key figure in the development of comparative psychology in the U.S. Working with invertebrates and rodents in his earlier studies, his interest later shifted to the study of primate behavior and intelligence, as reflected in his books *Almost Human* (1924) and *The Great Apes* (1929).

7 *Nagel*: Willibald Arnold Nagel (1870–1911), a German physiologist known for his work in the physiology of sensation and particularly for his study of the physiology of the eye.

8 *Lukas*: Franz Lukas (b. 1853), a Czech educator and philologist whose publications include a 1904 book on animal psychology (*Psychologie der Niedersten Tiere*).

9 *Royce's*: Josiah Royce (1855–1916), an American philosopher – a student of William James – and a proponent of a type of idealist philosophy called "absolute idealism," which, following German idealist philosophers G. W. F. Hegel and Friedrich Schelling, argued that reality consists of a single all-encompassing mind or consciousness.

Editorial Headnote

13. Charles Abram Ellwood, "The Origin of Society," *American Journal of Sociology* 15, no. 3 (November 1909): 394–404.

Part of the second generation of American sociologists, in the early decades of the twentieth century Charles Abram Ellwood (1873–1946) served as chair of the departments of sociology at the University of Missouri and Duke University, was a highly successful textbook author and the most translated American sociologist of the time, and generally considered a founding and leading figure of the discipline in the U.S. – none of which prevented him from plunging into relative obscurity during the postwar years. Ellwood's reformist sociological perspectives were interwoven with his Christian ethics, and he lamented what he perceived as American society's turn towards materialism and militarisation and American sociology's turn to a scientism devoid of any commitment to the cause of social and moral improvement. Increasingly estranged from the discipline he had helped establish, Ellwood's career, and soon after his life, ended with a bitter taste and a sense of failure at his inability to shepherd his vision of the purpose of sociology into a new generation.[1]

The question of the origins of society – and whether human society was to be understood in continuity with animal modes of association – which Ellwood addresses in his article for the pioneering *American Journal of Sociology*[2] had been a subject of long-standing debate. In his *Philosophy of Natural History*, eighteenth-century Scottish printer and naturalist William Smellie distinguished between "proper" and "improper societies," with the latter supposedly characterised by a mode of association without "a view to any joint operation."[3] Versions of this argument were reiterated by various nineteenth-century American writers such as Massachusetts clergyman Frederic Dan Huntington, who claimed that while it was true that "herds of cattle, flocks of birds, shoals of fishes" were clearly phenomena characterised by some principle of association, "society" could be found in none of them. "If mankind were associated only to eat, sleep, hunt, and secure animal protection, those brute terms would apply. . . . But, for society, there must be something more than an affinity for living together" – and this "more" was not only defined by the *how* but also by the *who* of the association in that "[t]he object of congregating must be the educating, maturing, completing, of a human nature; rational, affectional, immortal."[4]

Ellwood, on the other hand, did not hesitate to use the term "society" in reference to "association among animals" and even insisted that sociality was a feature of life as such in its more rudimentary forms. In many animal species, sociality resulted from the fact that association and cooperation among conspecifics was an integral or at least advantageous element in the basic life processes of food acquisition and reproduction and thus favored by natural selection. Human society, then, was neither divinely ordained nor the result of some compact or contract among humans in a hypothetical state of nature (Hobbesian or otherwise), who agreed to cede individual freedoms for security, protection, and cooperation. For Ellwood, there was no fundamental difference between human society and nonhuman societies, other than in the "forms and definiteness of the psychical interaction between individuals."

Notes

1 Stephen P. Turner, "Ellwood, Charles Abram," in *American National Biography*, ed. John A. Garraty and Mark C. Carnes, vol. 7 (New York: Oxford University Press, 1999), 458–60; or, more comprehensively, Turner, "A Life in the First Half-Century of Sociology: Charles Ellwood and the Division of Sociology," in *Sociology in America: A History*, ed. Craig J. Calhoun (Chicago: University of Chicago Press, 2007), 115–54.

2 The *American Journal of Sociology* was founded by Chicago sociologist Albion Small in 1895 as the first journal in its field and preceded the founding of the American Sociological Society (today American Sociological Association) by a full decade. See "The American Journal of Sociology Through Fifty Years," *American Journal of Sociology* 50, no. 6 (1945): 522–33.

3 William Smellie, *The Philosophy of Natural History*, vol. 1 (Edinburgh: Printed for the Heirs of Charles Elliott, 1790), 432.

4 Frederic D. Huntington, *Human Society: Its Providential Structure, Relations, and Offices* (New York: Robert Carter, 1860), 9–10.

13

THE ORIGIN OF SOCIETY (1909)

Charles A. Ellwood

Source: Charles Abram Ellwood, "The Origin of Society," *American Journal of Sociology* 15, no. 3 (November 1909): 394–404.

THE ORIGIN OF SOCIETY IN GENERAL

The origin of society in general, that is, of association among animals, and of human society in particular, can no longer be regarded as purely a speculative question. During the seventeenth and eighteenth centuries social philosophers gave so many and so varied answers to this question, from the supernatural to the contract theory, that it deservedly fell into disrepute. The advances of nineteenth-century science have made it evident, however, that the problem of the origin of society is no more insoluble than the problem of the origin of species. This is not saying, of course, that there remain no unexplained elements in the problem, or that there is general agreement among all sociologists upon this question. Life in general remains a mystery to science, and as long as it does the origin of association as a phase of the life-process must remain also to a certain extent a mystery.

Fundamentally the problem of the origin of society is a biological question. The psychological sociologist, in his discussion of the problem, needs only to point out that *the life-process is essentially social from the start;* that is, it involves from the first the interaction of individual organisms. This interaction, while in its lowest phases purely physical, gives rise in its higher stages to that psychical interaction which we call association or society.

Life is not, and cannot be, an affair of individual organisms. The processes of both nutrition and reproduction, in all higher forms of life, involve a necessary interdependence among organisms of the same species, which, except under unfavorable conditions, gives rise to group life and psychical interaction. Society is no more the result of the coming together of individuals developed in isolation than the multicellular organism is the result of the coming together of cells so developed. Society, that is, the psychical interaction of individuals, is an expression of the original and continuing unity of the life-process of the associating organisms. Looked at from the standpoint of the whole evolution of life, it is really the result of the breaking-up of the life-process into several relatively independent centers while the process itself remains a unity. The functional interdependence on the

DOI: 10.4324/9781003032717-14

psychical side which constitutes a group of organisms a society is a mark at once of their original unity in a common life-process and of the fact that they now constitute a higher, more complex unity. In this view, the social process is strictly a phase of the life-process, even in the biological sense.

The social process, then, grows spontaneously out of the life-process. It grows out of both of the fundamental phases of the life-process – the food-process and the reproductive process. The food-process, or the activities connected with nutrition, seems to act chiefly in a negative way upon the earliest beginnings of association. As a rule, organisms of one species remain together as long as food is abundant, and they scatter only when the conditions of nutrition become unfavorable.

The thing to be explained in the organic world is not the living together of large numbers of one species, but rather the scattering and separation of individuals. As has already been said, separation usually takes place on account of lack of food supply, while where food supply is abundant and sufficiently concentrated the individuals of a species remain together in large numbers. Now, where living forms remain in close proximity to each other they tend to take on functional interrelations both in the food-process and in the reproductive process. The conditions of food supply thus become the physical basis of the interrelations among organisms, interrelations which later become psychical. When the conditions of food supply become unfavorable, the tendency to scatter, moreover, may be overcome by new adaptations on the part of organisms which give rise to superior ways of co-operating, so that an adequate supply of food shall be assured. Or when scattering does take place it may be by bands, and those bands whose members co-operate best in finding a food supply would have the best chance of survival.

The control over the food-process is the matter of supreme concern both to the individual and to the species. Not only is a stable food supply necessary for the survival of the individual, but reproduction can take place only after nutrition has reached a certain height, and it tends to go on only where food supply is abundant. Now, control over the food-process can be more easily established by groups of co-operating individuals than by isolated individuals. Natural selection operates, therefore, from the first in favor of such groups, and toward the elimination of individuals living relatively isolated. It must especially favor those groups in which the interactions between individual units are quick and sure – that is, those groups in which the power of psychic interstimulation and response is fully established, and in which intelligent co-operation and orderly relations between individuals are highly developed. It is not an accident that the most successful and, in general, the higher animals live in groups with well-ordered relations and highly developed means of interstimulation and co-operation.

Thus does the collective control over the food-process, established primarily by natural selection, become the positive basis of social organization, so that it is possible even to say, in a rough way, that the social process is a function of the food-process. The goal, indeed, of much conscious social development seems to be the collective control of the food-process. Whether it is the only goal, or the highest goal, of social development will be considered later. It suffices to point out here that social organization and evolution present themselves, from one point of view, largely as a direct outgrowth of that fundamental phase of the life-process which we have called the food-process.

Defense against enemies may be regarded as the negative side of the food-process, since it is largely in the efforts to secure and maintain a food supply that the necessity of defense arises. That such defense can be much better undertaken by groups of individuals than by isolated individuals; and that natural selection, therefore, operates powerfully in this way alone to favor group life, have long been among the commonplaces of sociologists. The conflict of group with group in the struggle for the possession of the material means of subsistence has been one of the most important factors in social evolution, especially in the way of integrating groups. It is not our purpose, however, to discuss the workings of this factor in social evolution further than to point out that conflict, as a phase of the food-process, has contributed powerfully to the genesis and development of association or group life.

It is not, however, the food-process which has played the chief rôle in the genesis of association among animals. That honor belongs to the reproductive process, using that phrase in a broad way to cover all the activities connected with the birth and rearing of offspring. The birth and care of offspring are essential phases of the life-process, and at the same time are essentially social activities, since in all but the lowest forms of life they involve the co-operation of at least two individuals. Sexual reproduction, necessitating the interaction of two individuals, lays a positive foundation for association. It is, however, the production of immature or "child" forms which need prolonged and tender care on the part of one or both parents which gives rise to that most intimate form of association that we term the family, which produces and reproduces the social life from generation to generation and which becomes the basis, in large measure, of all later social organization. In the relationship of the mother to the child we have the beginnings of that sympathetic social life, of which the family has remained the highest type and which has become the conscious goal of civilized human society. Society in the sympathetic sense, then, had its beginnings in the family, that is, in the relation of the child-form to the mother-form.

The relationship of the child-form to the parent-form becomes more prolonged and increasingly important as organic evolution advances. While in the lower reaches of life the reproductive process is comparatively unimportant in its social results, in the higher animals with the prolongation of the period of immaturity and with the increasing necessity of the co-operation of both parents in the care of the young, it becomes supremely significant for the social life. While it is a law that the higher we ascend in the animal scale, the less energy is devoted to mere physical reproduction, it is equally a law that the higher we ascend in the animal scale the more energy is devoted to the care and rearing of the offspring that are born. The social results of the reproductive process become, therefore, increasingly rich, significant, and complex as we ascend in the scale of animal life. It is among the higher animals that the family as a form of association receives its highest development, and hitherto it has been among the most highly civilized peoples that the family as a human institution has been held in highest regard and most safeguarded in custom and in law.

It is not, therefore, too much to say that the social process is a function of the reproductive process quite as much as it is a function of the food-process; that the social order exists to safeguard the birth and upbringing of each new generation quite as much as to assure an adequate supply of material goods to those already existing. Of course, these two phases of the social process are supplementary and should not be set in opposition to each other. They would not need to be distinguished, were there not some who talk as if the only function of the social life were to secure for all an adequate supply of material goods. Certain it is that all forms of social life, from the ants and bees to man, and in the human world from savage to civilized, have been determined from considerations of reproduction quite as much as from considerations of nutrition. The goal of social development is, therefore, quite as much control over the reproductive process as control over the food-process. The child is not only the center of the family life, but of the whole social system as well. The child's heredity, birth, care, and education are the supreme concern of church and state as well as of the home, and the sooner this is recognized the better.

If the general forces at work in the genesis of association or group life are now clear, it remains only to say a word about the social character of the individual mind; that is, how consciousness comes to be the chief connecting link between individuals living in association. As far back as we can go in mental evolution the psychic elements of life are a chief means of binding individuals of the same species together. Instincts, emotions, and sensations of one individual organism often seem made to fit into corresponding mental processes of another organism; and varied means of interstimulation and response are developed. The mind seems to be social in its nature from the start, and to be at once a social product and a social instrument.

The reason for this is now clear. Consciousness is concerned with the mediation of the activities of the life-process, particularly those of the food-process. But the life-process of the individual is only a part of the larger life-process of the group to which he belongs. The procuring of food and the protection against enemies, as we have seen, are activities which can be more successfully carried on by the group than by the individual. But consciousness is concerned with the mediation of these life activities. If they are carried on by groups it is evident that the only way the mind can control them is through some form of psychic interconnection between the individuals of the group. Hence have arisen the various forms of psychical interaction (interstimulation and response) between individuals. These forms of psychical interaction, in man at least, are so perfect that intelligence controls collective action almost as easily as individual action. Thus the social character of mind is an expression of the fact that it has to do with the mediation of a process which is carried on by several co-operating individual units; while society, the psychical interrelations of these individuals, means that there is one common process of living carried on by these co-operating units on the psychic plane, that is, on the plane of interstimulation and response.

THE ORIGIN OF HUMAN SOCIETY

The position already implied is that the processes involved in human association are fundamentally the same as in animal association; in other words, that animal society is the precursor of human society, and that, strictly speaking, human society is but a form of animal society. Human society is, however, so different from animal society that it is considered by many to be *sui generis.* But the whole difference between the two, it can readily be shown, is in the forms and definiteness of the psychical interaction between individuals. What especially distinguishes human society from animal groups is the possession of articulate language. It is this which makes possible the communication of definite ideas, giving a far greater degree of definiteness to the whole process of social interaction and making possible among human beings many higher forms of co-operation. Articulate speech, of course, rests in some degree upon the power of forming abstract or general ideas, though it in turn reacts to develop that power. Upon these two great differences between man and the other animals – articulate speech and the power of abstract thought – rest the chief differences between animal and human society; for the other great distinctive marks of human society, such as the rationality and self-consciousness of its individual members, religion, and government, all go back to, or are intimately associated with, language and the power of abstract thought.

If what has been said is true, then human society must be regarded as an inheritance from man's prehuman progenitors, and as a form of animal society. Even many of the forms of human association were doubtless fixed in the sub-human stage. This is notably true of man's family life, which in its essential features, as Westermarck[1] and others have shown, must be regarded as an inheritance from man's apelike progenitors. It is also true of such a form of association as leader and follower, for the phenomena of leadership are found among many of the higher animals.

In a word, human society rests upon instincts established by natural selection during the long prehuman stage of man's evolution. These instincts were the basis of all the primitive forms of association among men, and the addition to these of the intellectual elements of language, abstract ideas, self-consciousness, and reason is what gave rise to the peculiar products of human social evolution, human institutions, and civilization.

The origin of these intellectual elements which have given a peculiar color and form, so to speak, to human association we cannot here discuss except to say that they are themselves largely social products. Language is manifestly a social product, and the fact that man is the only speaking animal is correlated with the fact that he is pre-eminently the social animal. In the same way, the power of abstract thought and of syllogistic reasoning[2] may be shown to depend largely upon language and other traits developed through association. Even self-consciousness itself, the consciousness of the unity and continuity of our mental life, which many make the distinctive mark of human society, is probably an outcome of association. It certainly depends for its development in the child largely upon language and the general give-and-take of the social life. All this, of course, is equivalent to saying that the differences between animal and human society are due to the natural social evolution of the human species; that the causes of these differences are to be sought in human social life itself, and not outside.

This is not saying, of course, that there may not be instincts peculiar to man as an animal which account in part for the differences between animal and human society. But it is saying that these peculiar human instincts are not what give human society its distinctive character, but rather the intellectual elements; and that these instincts have evolved, and all man's instincts been modified, under the influence of a social life in which intellectual elements were powerful. Thus are harmonized the instinctive and the intellectual elements in human society.

The family life of man, as the primary form of human association, will serve to illustrate these points. Though man's family life in its essentials is undoubtedly an inheritance from his prehuman precursor, yet one is struck at once by the vast differences between the family life of man and that of the higher animals nearest him. There is, for example, in the human species no pairing season, little tendency to natural ornament during the period of courtship, but a strong tendency to artificial adornment, while there seems to be an instinct against incest, preventing close inbreeding. These differences may be perhaps set down to a difference in instincts between man and the higher animals. On the other hand, there are many differences which cannot be so explained, such as the fact that the indorsement of society is almost invariably sought among human beings before the establishment of a new family, usually through the forms of a religious marriage ceremony; that there exists a feeling of modesty regarding matters of sex; and that chastity is enforced, on married women at least, among all peoples. While these peculiar traits of human family life may perhaps in part be traced to peculiar human instincts, yet the element of self-consciousness in each of them is so large and so manifest that they may be safely ascribed largely to man's intellectual nature. Thus the human family life illustrates both the instinctive origin of human association and its modification through intellectual elements which have caused it to vary widely from the primitive animal type.

Here must be noticed the influence of the prolongation of human infancy upon human social life. This purely biological fact, whose importance John Fiske[3] was the first to call attention to, has had a profound influence on both the instinctive and intellectual elements in human association, and especially on human family life. We have already noted how the prolongation of the period of immaturity of offspring affects social evolution in general, cementing the union between the parents and giving opportunity for the development of the sympathetic instincts and emotions within the family group. It is no doubt largely due to prolonged human infancy, therefore, that we have regularly in human society a permanent union between the parents lasting throughout life; permanent sympathetic relations between all members of a family group, giving rise to the sentiment of blood kinship; and a high development of sympathetic feeling and altruism in human society generally. It is, however, often overlooked that the prolonged period of immaturity in man, besides cementing the human family group and generating altruism in an instinctive way, gives opportunity for the intellectual elements in human association to assert their influence. It is prolonged immaturity which makes education possible, and gives opportunity for social tradition to mold each individual in conformity with the habits of his social group. Language could hardly be transmitted, and could not be developed and perfected without prolonged immaturity. And so with every other spiritual possession of humanity. Abstract thought, religion, government, and moral ideals could hardly effectively mold individual conduct or influence the social organization, were it not for the period of relatively prolonged plastic immaturity through which every individual passes. Upon this biological circumstance depend, therefore, many of the striking features of human social life, especially the influence of intellectual elements, that is, plasticity and ultimately the capacity for social progress itself.

Other peculiar features of human social life, which by some are held to be so peculiar as to make human society in a class by itself and not comparable with animal groups, may now be quickly disposed of. It is said that man transforms the environment while the environment transforms the animal. While the contrast in such absolute terms is not justifiable, yet it must be admitted that man's growing mastery over physical nature is one of the most striking facts of human social life. But it is evident that it is but an outgrowth of man's power of abstract thought together with that vast cooperation which human science and art imply. It is a secondary, then, rather than a primary difference between human and animal social life. Again, the existence of a conscious social morality in human groups has been claimed to be an irreducible difference between them and animal groups. But even Aristotle perceived that this was due to the fact that human groups possess language and so social tradition, and, we may add, the power of abstract thought to form ideals. Organized government is a distinctive feature of human societies, although not all possess it. But organized government undoubtedly rests upon the same foundations as social morality, with perhaps an even larger rational and deliberative element. Finally, religion is a distinctive feature of all human groups whatsoever, but it is probably a product of the interaction of man's self-consciousness and reason with his instinctive life.

To sum up: We may conclude, then, that the social development which we find in humanity is in principle the same as the social development which we find in the animals below man; that the origin of human society is in the instincts established by natural selection long before the human stage was reached, though the development of human society has been largely modified by intellectual elements. Though these intellectual elements are important, human society is not in any sense an intellectual construction due to the perception of the utilities of association. It is not a contract, as was once thought, which can be made over to suit the pleasure of the parties thereto; neither is it a machine of the gods which man cannot modify. Human society is modifiable in the same sense and in the same degree in which human nature is modifiable. While social organization, customs, and institutions rest fundamentally upon instincts which have grown out of the necessities of the life-process, these instincts and the habits which grow out of them are modifiable by intellectual elements, especially in the young. Education is the only sure means, and I believe, the only safe means, of social reorganization.

Editorial Endnotes

1 *Westermarck*: Edvard Alexander Westermarck (1862–1939), a Finnish sociologist and philosopher mostly known for his studies on the histories of ("primitive") marriage and moral ideas.

2 *syllogistic reasoning*: a syllogism is form of deductive argument that consists of two premises and a conclusion ("All animals are mortal; wombats are animals; therefore, wombats are mortal").

3 *John Fiske*: John Fiske (1842–1901) was an American philosopher, historian, and populariser of evolutionary theory in the U.S. who was strongly influenced by the work of Herbert Spencer (see J. Fiske, Vol. 2).

Editorial Headnote

14. Ira Woods Howerth, "The Great War and the Instinct of the Herd," *International Journal of Ethics* 29, no. 2 (1919): 174–87.

Through the 1890s, Ira Woods Howerth (1860–1938) studied at Harvard and the University of Chicago, where he completed his PhD in 1898 and then became the head of its sociology department. He later taught at the University of California, Berkeley and the State Normal School of Colorado (today University of Northern Colorado), publishing extensively in the areas of education and sociology.[1] A peace advocate and, like Ellwood, an opponent of American militarisation, Howerth's piece on the "Great War" in the *International Journal of Ethics* – founded in 1890 by leaders of the Ethical Movement and continuing today as *Ethics*[2] – is grounded in his belief that, in all its unprecedented brutality, World War I was only the most recent expression of a much greater "age-long conflict" of a "war of civilization against war" – a conflict in which, Howerth believed, the forces striving for "universal peace" were ultimately bound to emerge victorious.[3]

Howerth's concern with the social psychology of (mass) conflict at the core of his 1919 article is already evident in his earlier essays "Patriotism, Instinctive and Intelligent" and "War and the Progress of Society," the latter of which conceptualises war as a "socially unconscious phenomenon" akin to "earthquakes, floods, famine and pestilence."[4] Howerth was not alone in employing the concept of the herd instinct as an analytic lens to make sense of the cataclysm of WWI. One of the early theorists of the herd instinct was English neurosurgeon Wilfred Trotter in his two-part essay "Herd Instinct and Its Bearing on the Psychology of Civilized Man," published in the *Sociological Review* in 1908/09 and his book *Instincts of the Herd in Peace and War* (1916). Both Trotter and Howerth were heavily influenced by French sociologist Gustave Le Bon's work on crowd psychology and his claim that crowds had to be understood as psychological entities in their own right whose behavior was dominated by a collective unconscious that temporarily suspended the sense of self, as well as the reasoning capacities and moral beliefs, of its individual members.

Howerth's discussion begins from the fact of the intensely "gregarious" nature of human beings, which they shared with other herd animals. For Howerth, the modern nation, too, was a herd – one that perhaps differed from other herds in scale and complexity but otherwise operated on the basis of the same principles and underlying tendencies. As a "conflict of herds," warfare between nation states was the result of the "overwhelming power of the herd feeling" – specifically of the herd's "tendency to unite" and its "susceptibility to suggestion" and "docility in the hands of leaders."

Notes

1 *Who Was Who in America*, vol. 1 (Chicago: Marquis Who's Who, 1943), 597.

2 The beginnings of the Ethical Movement date back to the founding of the New York Society for Ethical Culture in 1876 by German-born Felix Adler, a graduate of Columbia University who would return to his *alma mater* as a Professor for Political and Social Ethics in 1902. The Ethical Movement sought to promote ethical living based on humanistic values rather than religious doctrine, emphasizing the cultivation of personal morality and social or communal responsibility. The founding of the New York Society led to the creation of Ethical Societies in other U.S. cities like Chicago, Boston, and Philadelphia and to the internationalization of the movement. See *The Fiftieth Anniversary of the Ethical Movement, 1876–1926* (New York: D. Appleton, 1926); Howard B. Radest, *Felix Adler: An Ethical Culture*. New York: Peter Lang, 1998.

3 "War of Civilization on War Is Described. Age-Long Struggle Eclipses European Strife, Says Peace Advocate," *San Jose Mercury News*, October 29, 1914.

4 Ira Woods Howerth, "War and the Progress of Society," *Popular Science Monthly* 87 (August 1915): 195–99, here 199; Howerth, "Patriotism, Instinctive and Intelligent," *Educational Review* 44 (June 1912): 13–24.

14

THE GREAT WAR AND THE INSTINCT OF THE HERD (1919)

Ira W. Howerth

Source: Ira Woods Howerth, "The Great War and the Instinct of the Herd," *International Journal of Ethics* 29, no. 2 (1919): 174–87.

DARWIN said, "Judging from the habits of savages and of the greater number of the Quadrumana,[1] primeval men, and even their ape-like progenitors, probably lived in society."[a] But the relative weakness of early man, the consequent necessity and obvious advantage of co-operation in defence against his enemies, his success in the struggle for existence as shown by his wide diffusion, as well as the social disposition of the animals nearest him in the scale of life, make it almost a certainty that primeval men were social, that they lived in groups, that they possessed the herd instinct. While there is a difference of opinion, as McDougall[2] remarks, as to the exact form of early society, as to whether it was a promiscuous horde or a group made up chiefly of blood relations, something after the nature of a patriarchal family, anthropologists are practically agreed that primitive man was gregarious in his habits.[b]

While we read without surprise that primitive man was gregarious, we are likely to forget that modern man, although many of his acts have lost their fixed and instinctive character, or been replaced by acts of will, is quite as gregarious as his primitive ancestor. This is clearly shown by his disinclination to solitude, by his normal forms of recreation, and by the disposition of men to herd together in towns and cities. Solitary confinement is everywhere regarded as a mode of punishment peculiarly severe, and is looked upon by many as a form of torture so peculiarly dreadful that it should be prescribed only in exceptional cases, and in civilized countries not at all. "Misery loves company"; so also does happiness. It is a sociological fact, then, confirmed by prison discipline, as well as by bachelorhood, that it is not good for man to be alone.

As to our modes of recreation, everyone must have observed that with most people the one essential condition of having a good time is to be one of a crowd. People flock together at the seashore, in public parks, and in places of entertainment and amusement. Many go to church, even, to see and be seen, to satisfy their herd instinct, rather than to profit by religious instruction. Again, the extraordinary

DOI: 10.4324/9781003032717-15

growth of cities, not only in this country but in all civilized lands, is due not alone to economic motives but also to the blind instinct which drives men to seek the company of their fellows. Man, then, is a gregarious animal, as gregarious as the horse, the elephant, the ox, the buffalo, the sheep, the deer, the bee, or the ant.

Now the satisfaction of this gregarious or herd instinct in man has resulted in the formation of many and various social groups: families, clans, tribes, states, nations; and, within the nation, parties, professions, denominations, clubs, sects, classes, and the like. But the largest social unit thus far definitely formed is the nation. In times of peace its herdlike qualities are not conspicuously manifested, but in war, when the bonds of instinctive feeling woven in the loom of biological and social history, and "The mystic cord of memory stretching from every battlefield and patriot grave to every living heart and hearthstone,"[3] as Lincoln expressed it, – when these draw the people together, herd phenomena begin plainly to manifest themselves. We call them manifestations of patriotism, but patriotism is at bottom an instinctive herd emotion. It is "the emotion of his national crowd in the heart of the individual citizen."[4]

A nation, then, is a herd. This is not a fanciful analogy; it is a sociological fact, none the less real because in normal times it is only dimly revealed. A drove of cattle grazing peacefully on a plain is none the less a herd, although distinctive herd phenomena do not then conspicuously appear. Of course in calling a nation a herd we use the word in a purely scientific sense and with none of the disparagement that is implied by the phrase "the common herd."

Now in the recent world-war situation we had essentially a conflict of herds. We should therefore have expected that in such a conflict the usual and primary characteristics of the herd would be manifested. Let us turn to a brief consideration of a few of these characteristics.

First, then, the herd unites in the presence of danger. This is a well-known fact in the case of animals. Cattle or horses may be grazing peacefully apart, but at the sign of danger they draw together. Wolves fighting among themselves cease when threatened by a common foe; and I have heard it said that the same effect is produced by outside interference in a family quarrel. With nations the effect is the same. The North in 1861 was unified as by an electric spark when Fort Sumter[5] was fired upon. [. . .]

A similar effect was produced in Russia in 1812 by the invasion of the French army. "From the moment when Napoleon had crossed our frontier," wrote the Empress Elizabeth,[6] "it was as though an electric spark had spread through all Russia; and if the immensity of its area had made it possible for the news to penetrate simultaneously to every comer of the empire, a cry of indignation would have arisen so terrible that I believe it would have resounded to every comer of the earth."[c] The psychologic effect produced upon a nation by a common danger is so well known that more than one statesman has aimed to create it by prescribing a foreign war as a certain remedy for internal dissensions. Bismarck had it in mind when by altering a telegram in 1870 he provoked the French to declare war against Germany.[7] Seward's[8] famous letter to Lincoln in 1861, advising war with Spain and France in order to unite the North and the South, is another example out of many. He wrote in "Some Thoughts for the President's Consideration, April 1, 1861," "My system is built upon this *idea* as a ruling one, namely, that we must change the question before the public from one upon slavery, or about slavery, for a question upon union or disunion. In other words, from what would be regarded as a party question, to one of *Patriotism* or *Union*. . . . I would demand explanation from Spain and France, categorically, at once. I would seek explanations from Great Britain and Russia, and send agents into Canada, Mexico, and Central America, to rouse a vigorous continental spirit of independence on this continent against European intervention. And, if satisfactory explanations are not received from Spain and France, would convene Congress and declare war against them."[d] Thus he planned to divert the thoughts of the people from internal dissensions, and unite them in a common feeling against a foreign foe, that is, to utilize for political reasons the instinct of the herd. The herd tends to unite in the presence of danger. This is as true of the nation as it is true of a drove of cattle, a flock of sheep, or a pack of wolves.

Now, observe that the unity thus occasioned is a unity arising from feeling and not from thought. It is instinctive, not rational. This is plainly the case with animals. They do not reason, but they unite just the same. With man reason may be and is at times and in particular cases a check upon union; but with the multitude reflection quickly succumbs to a sense of common danger, or of injured national honor, dignity or prestige; and unity is effected, and effected quickly because feeling is contagious. This instinctive disposition to unite in the presence of danger is obviously an advantage in the struggle for existence. Without it a group could not survive in the presence of united aggression. A nation that stopped to reason and argue itself to a rational basis of union would be defeated before it began to fight.

We are not concerned at present, however, with the value of the herd instinct or the dangers arising from it. We wish merely to point out that it sufficiently explains the unity existing in the countries recently at war. The people of a country may be expected to unite against a common foe no matter what the national cause may be. The union is instinctive, spontaneous, a union of feeling and not of thought. Hence the unanimity of a people at war has nothing whatever to do with the justice or righteousness of their cause. This explains why the Germans were united in unrighteousness. It explains also why the pacific principles of the German social democracy melted away as soon as war was declared. These principles and doctrines were rational formulations, matters of the head, but the national sentiment expressed in *Deutschland ueber Alles*[9] was a matter of the heart, and the heart is older than the head. It has been well said that emotionally we are hundreds of thousands of years old, while rationally we are embryos.

The prevalent assumption, then, that the common people of Germany were driven into war by the ruling powers, or that there was during the war any essential difference in Germany between the national feeling of Kaiser and cobbler, was without foundation. The German Emperor expressed the real situation when he said, at the opening of the Reichstag immediately after war was declared, "We are all Germans now."[10] The refrain in the German "Hymn of Hate,"[11] "we love as one, we hate as one" was the expression of a psychological fact based on the herd instinct. A recognition of the presence and power of this instinct would have prevented the expectation of an early revolution in Germany.

Before leaving this part of the subject we may observe that because of the instinctive reaction of the people of a country against the circumstances which provoke war, as, for instance, a national insult, a popular referendum of the question of peace or war, however desirable it might be on other grounds, would not be a sure mode of preventing war. In case of a real or fancied lesion of the national honor, dignity, or prestige, the people are even more likely to "lose their heads," that is, yield to the herd instinct, than are those in authority. Hence democracies are as prone to war as monarchies. The "king business" should be broken up, but war will not necessarily disappear with kings.

Another mistaken conception, entertained by some before the war, on account of their failure to recognize the reality and power of the herd instinct, was the idea that monied men, particularly those in control of the great banking interests, would not permit another war to take place. They had too much to lose. We see now how illfounded was this view. A knowledge of the herd instinct, and of the overwhelming power of herd feeling when once incited would have prevented that mistake.

So much, then, for the first characteristic of the herd, its tendency to unite. A second characteristic is its susceptibility to suggestion. This susceptibility is obviously a fact with respect to a herd of animals; it is peculiarly and conspicuously so with respect to a human herd, especially a mob. In a mob any suggestions of fear, anger, enthusiasm, elation, depression, etc., sweep through it almost instantaneously, and bring forth an immediate response. Hence, the mob is notoriously fickle, sentimental, unreasoning; now brave, now cringing, now cruel, now kind. Specialists in crowd psychology have pointed out, what anyone may easily observe, that a man in a crowd is very different from the same man when he is an isolated individual. His conscious individuality vanishes in the unconscious personality of the crowd. The crowd, we are told, has a collective mind and a collective consciousness. It is dominated by unconscious elements; it is credulous and short sighted. Says Le Bon, "By reason of the extreme sensibility of crowds, their sentiments, good or bad, are always exaggerated. This exaggeration increases still further in times of revolution. The least excitement will then lead the multitude to act with the utmost fury."[e]

The truth of this observation is well illustrated by the doings of crowds in the French Revolution. With Liberty, Equality and Fraternity on their lips Frenchmen, under the influence of the crowd spirit, indulged in orgies of blood-shed and cruelty no less revolting than the sack of Belgium or the murder of Edith Cavell.[12] Thousands and thousands of innocent people were massacred; men and women, even old people, were burned alive; women and children were violated and afterwards murdered; tender babes were torn from their mothers' arms and tossed from bayonet to bayonet; men were compelled to dig their own graves and suffer themselves to be buried alive. No wonder Madame Roland[13] cried out, on her way to the guillotine: "Liberty, liberty, what crimes are committed in thy name!"

Now in all countries, in times of peace as well as in times of war, we may expect that there will be occasional manifestations of the crowd spirit. Citizens who are ordinarily peaceable and inoffensive will now and then assemble in a crowd and, seized with the crowd spirit, commit the most fiendish barbarities, – condemn an accused man without trial, for instance, and perhaps burn him at the stake; thus enacting in civilized communities the identical barbarity which we are accustomed to think of as peculiar to savage tribes.

Because of the stupidity, passion, and consequent danger of crowds every nation should endeavor through education to fortify its citizens against the contagion of the crowd spirit, and those in authority should do everything possible to prevent the formation of crowds. President Wilson displayed a high quality of leadership in his direct appeal to the people of this country, as soon as patriotism began to manifest the spirit of the mob, to avoid such manifestations as unworthy of a free and intelligent people.

But in Germany it so happened that the mode of making war thought to be most effective was directly provocative of the crowd spirit. When the German army laid waste every foot of land over which it retreated, destroyed every habitation; when it shot down in cold blood its hostages, or even all the inhabitants of a street or village in which someone had committed a hostile act, it acted in strict accordance with the military principles laid down by the German Staff. *Schrecklichkeit* was a deliberate policy adopted by that staff, and with professedly humane motives. "Terrorism," said General von Hartman,[14] "is seen to be a relatively gentle procedure, useful to keep in a state of obedience the masses of the people."[f] But when soldiers proceed to carry out that policy, when the butchery begins, then a company of soldiers becomes a mob fired by the mob spirit, and there is no limit to the atrocities it will commit. "Nothing could be done against the excited crowd," said a German soldier with some small remnant of pity, "for at such times they are not men but beasts." "One gets an idea of the madness of our soldiers," says another, "when one sees the demolished villages. Not a single house intact. . . . Sheer vandalism was present as well as just anger." Carrier,[15] a wholesale murderer of the French Revolution, declared that he took an intense joy in seeing his victims suffer; that he never laughed so much or experienced such pleasure as in watching the dying grimaces of the priests he killed. And so a German soldier notes in his diary, "It is for me a mad joy when we can revenge ourselves on these rascally Belgian and French priests."[g] Accounts of the atrocities of the German army are incredible only to those who are unacquainted with crowd psychology, and who do not perceive the relation between the policy of "frightfulness" and the development of the crowd spirit.

But while the herd is always capable under excitement of extreme manifestations of the mob spirit, the general truth with which we are here concerned is only that it is highly suggestible. It is uncritical in its attitude and accepts the most improbable statements. This fact was well illustrated by happenings in all of the warring countries. In Germany, for instance, as soon as war was declared, stories of the poisoning of rivers, the infection of water supplies by cholera microbes, and of the omnipresence of foreign spies, were generally circulated and generally believed. Circumstantial accounts of the assassination of more or less distinguished German citizens were published, and wild excitement aroused thereby, when these same citizens were alive and well. The German newspapers of the time show that the patriotic delirium and spy mania of Germany worked great harm even to the Germans themselves. It was not greatly different in other countries.

So much for herd suggestibility. A third characteristic of the herd is its docility in the hands of leaders. The herd follows the leader. So does the crowd. In a mob a leader is almost sure to appear. So in times of stress and excitement nations look to their leaders, and accept as leaders only those who sum up in themselves the fundamental feelings and spirit of the group. Because they are powerful centers of suggestion these leaders exercise an uncanny influence. They determine in large measure the action of the people. Their responsibility is therefore great. Their power to lead is also a power to mislead! and if in times like these leaders do mislead, they will deserve all the condemnation they are likely to receive from impartial history. In the light of the susceptibility of a nation to suggestion, and its disposition almost blindly to follow its leaders, it is foolish to say, as some do, that nobody was responsible for the great war. We were all responsible more or less, perhaps unconsciously. But the responsibility lay primarily with those who willed it, and immediately with those who, because of their position, power, and prestige, were able to make their will effective.

These, then, are a few of the characteristics of the herd, whether it consists of animals or of men: it unites, it is highly susceptible to suggestion, it is controlled by leaders.

Now, in these respects, and in others, all herds, all nations, are alike. Still there are wide differences among animal herds and there are also wide differences among nations. A flock of sheep, a hive of bees, and a pack of wolves are alike in some respects but their differences are conspicuous. The same is true of nations. They resemble each other in some respects but in others they are as widely separated as the poles. How are we to account for these wide differences among nationalities, particularly among the recently opposing groups?

We cannot here enter upon a comparison of the score of nations that were engaged in the war, or even of the various animal herds. Such a comparison would be interesting and suggestive. A comparison of nations would have to be based, not upon superficial manifestations, but upon the particular ideas and ideals which in them have become common sentiments, specially ingrained in the people, and thus able to control national action. These ideals, sentiments, and interests constitute the character of a people. "Behind the institutions, arts, beliefs, and political upheavals of each people, lie certain moral, and intellectual characteristics that determine its evolution. . . . This aggregate of psychological elements observed in all the individuals of a race constitutes what may rightly be called the national character."[h] The fundamental differences between nations, then, are differences in character, just as their fundamental resemblances are due to instinctive reactions.

Now if Germany as a nation – whatever may be said of its citizens individually – if the German herd displayed in the war a relatively low order of morality, the fact must be attributed to the sentiments and ideas that have become characteristic and dominant among the German people. What are these peculiar psychological elements? In a word they are the sentiments and ideas of militarism. Militarism has been the curse of Germany as it will be the curse of any other nation that adopts it.

We have now seen that because of the gregarious nature of man all the social units thus far formed, including the nation, are of the nature of herds, and manifest the characteristic herd phenomena; and that many of the conditions and events of the Great War can be understood or explained only on the basis of the herd instinct. It goes without saying that this instinct has served a valuable purpose in social evolution. Without it social cohesion would fail a nation at critical times, and unity of action, without a high development of the social consciousness, would be impossible. A significant tendency is revealed, also, in the group-forming influence of the herd instinct.

That tendency is toward the formation of larger and larger social groups. In the great world war we had the incipient formation of two groups of humanity larger than the nation, two overherds as it were, consisting of the nations allied on either side, and in which the primary phenomena of the herd began to appear. It is improbable that the victorious nations associated in the war, which fought together against a common foe, which thought and acted in unison in matters relating to war, which are more or less conscious of a common national purpose, will ever return to their former state of relative isolation and independence. They are more likely to remain united, to "form a more perfect union" to develop a distinctive character of its own. If this should be the case it will point to the realization of the dream of world union. [. . .]

But, be that as it may, and admitting the value of this instinct as a factor in group survival, it forms no satisfactory basis for a distant unity of nations, or for the unity of a nation of to-day. After all is said, the herd instinct remains an instinct, and it can play no higher role with respect to a nation than is played by animal instinct in the biological world. Among animals instinctive reactions often result in wonderful adaptations. But while instinct is itself a product of evolution, it does not produce evolution; it is a passive element; it creates nothing; the more nearly perfect it is, the more likely it is to lead to the destruction of the animal species possessing it if the conditions under which it was developed undergo a change. The very essence of an instinct, as Darwin remarked, is that it is followed independently of reason. Intelligence alone is dynamic. It is the factor by which instincts are modified, and instinctive impulses inhibited, restrained, and directed. It is therefore the only safe reliance of a nation. Knowledge alone can make the world safe for democracy; only the truth can make us free.

Source Endnotes

a *Descent of Man*, Pt. I, Ch. 4, p. 78.
b McDougall, W., *Social Psychology*, p. 85.
c Quoted by Le Bon, *The Psychology of Revolution*, p. 107.
d Abraham Lincoln, a History, Nicolay and Hay, *Cent. Mag.*, Vol. 35, p. 615.
e The Psychology of Revolution, p. 102–3.
f Militärische Notwendigkeit und Humanität, in *Deutsche Rundschau*, Vol. 13, p. 462.
g Cited by T. de Wyzewa in *Revue des deux Mondes*, May 3, 1915.
h Le Bon, G., *Psychology of Peoples*, pp. 5–6.

Editorial Endnotes

1 *Quadrumana*: a historical term for "four-handed" primates capable of grasping with their feet. The term was introduced by German physician Johann Friedrich Blumenbach to distinguish monkeys, apes, and lemurs from *Homo sapiens*, the sole representative of the class "Bimana."
2 *McDougall*: William McDougall (1871–1938) was a British psychologist. Among his contributions was the development of a theory of instincts and pioneering work in the field of social psychology. His *Introduction to Social Psychology* was reprinted over twenty times in the three decades following the publication of the first edition in 1908.
3 *"The mystic cord . . . heart and hearth-stone"*: this is from Abraham Lincoln's first inaugural address at the Capitol on March 4, 1861.
4 "*the emotion . . . heart of the individual citizen*": from the first chapter of English explorer, historian, and politician William Martin Conway's (1856–1937) book *The Crowd in Peace and War* (1915).
5 *Fort Sumter*: the surrender of Fort Sumter – a sea fort located in Charleston Harbor, South Carolina, that was still incomplete in 1861 – to Confederate forces on April 13, 1861, marked the beginning of the Civil War.
6 *Empress Elizabeth*: the daughter of Tsar Peter the Great and Catherine I, Elizabeth Petrovna (1709–1762) was the Empress of Russia from 1741 to 1762.
7 *Bismarck . . . provoked the French to declare war against Germany*: Otto von Bismarck (1815–1898), a Prussian statesman who, from 1871 to 1890, served as the first Chancellor of the German Empire, played a decisive role in provoking the Second French Empire into a declaration of war, leading to the outbreak of the Franco-Prussian War (1870–1871). The telegram mentioned by Howerth is known as the Ems Dispatch, which was made public on July 13, 1870. It described an encounter between King Wilhelm I of Prussia and Vincent Benedetti, the French ambassador to Prussia, and was altered by Bismarck in a way that antagonised the French. The declaration of war followed shortly after on July 19.
8 *Seward's*: William Henry Seward (1801–1872) was an American politician who served as Secretary of State under Abraham Lincoln. The president rejected the recommendations contained in Seward's overly bold memorandum of April 1, 1861.
9 *Deutschland ueber Alles*: "Germany above all," the first line of German poet A. H. Hoffmann von Fallersleben's (1798–1874) "Das Lied der Deutschen" ("The Song of the Germans," written in 1841).
10 *The German Emperor . . . "We are all Germans now"*: this likely refers to Kaiser Wilhelm II's declaration in a famous speech to the German parliament (*Reichstag*) on August 4, 1914, "I no longer know [political] parties, I know only Germans" ("Ich kenne keine Parteien mehr, ich kenne nur Deutsche"). He used similar rhetoric in one of his "balcony speeches" (referring to the balcony of Berlin's Royal Palace), declaring

on August 1 that "I no longer know any parties or any confessions; today we are all German brothers and only German brothers."

11 *"Hymn of Hate"*: the "Song of Hate against England" ("Hassgesang gegen England") was written in 1914 by German-Jewish poet Ernst Lissauer (1882–1937).

12 *the sack of Belgium or the murder of Edith Cavell*: the sack of Belgium refers to a series of atrocities committed by German troops during the invasion of the neutral country in August 1914 and the following years of occupation, which include the infamous burning and pillaging of the city of Leuven. Edith Louisa Cavell (1865–1915) was a British nurse who worked with the Belgian resistance to help French and British soldiers trapped behind enemy lines escape from occupied Belgium. She was arrested by German authorities in August 1915 and subsequently sentenced to death for "war treason."

13 *Madame Roland*: Marie-Jeanne Roland (1754–1793) was a philosopher, *salonnière*, and prominent figure of the French Revolution. As a member of the moderate Girondin faction, Roland was arrested during the lead-up to the Reign of Terror (September 1793 to July 1794) and sentenced to death after several months of imprisonment and a brief trial.

14 *General von Hartman*: Julius von Hartmann (1817–1878) was a Prussian cavalry general who fought in the Austro-Prussian and Franco-Prussian Wars (1866 and 1870–71, respectively). In 1877/78, Hartmann published a series of essays on "military necessity" that justified "terrorism" as a legitimate military principle.

15 *Carrier*: Jean-Baptiste Carrier (1756–1794) was a Jacobin responsible for the mass murder (through drowning in the Loire) of thousands of civilians during the Reign of Terror.

INDEX

Note: Numbers in *italics* indicate a figure. Numbers followed by n followed by a number (ex: 12n3) indicate a footnote by editors. Numbers followed by alphabet letters (ex: 12a, 12b) indicate a footnote made by the original author of a reprinted article.